Genetics and Fish Breeding

CHAPMAN & HALL FISH AND FISHERIES SERIES

Among the fishes, a remarkably wide range of fascinating biological adaptations to diverse habitats has evolved. Moreover, fisheries are of considerable importance in providing human food and economic benefits. Rational exploitation and management of our global stocks of fishes must rely upon a detailed and precise insight of the interaction of fish biology with human activities.

The *Chapman and Hall Fish and Fisheries Series* aims to present authoritative and timely reviews which focus on important and specific aspects of the biology, ecology, taxonomy, physiology, behaviour, management and conservation of fish and fisheries. Each volume will cover a wide but unified field with themes in both pure and applied fish biology. Although volumes will outline and put in perspective current research frontiers, the intention is to provide a synthesis accessible and useful to both experts and non-specialists alike. Consequently, most volumes will be of interest to a broad spectrum of research workers in biology, zoology, ecology and physiology, but an additional aim is for the books to encompass themes accessible to a non-specialist readers ranging from undergraduates and postgraduates to those with an interest in industrial and commercial aspects of fish and fisheries.

Applied topics will embrace synopses of fishery issues which will appeal to a wide audience of fishery scientists, aquaculturists, economists, geographers, and managers in the fishing industry. The series will also contain practical guides to fishery survey and analysis methods and global reviews of particular types of fisheries.

Books already published and forthcoming are listed below. The Publisher and Series Editor would be glad to discuss ideas for new volumes in the series...

Available titles

1. **Ecology of Teleost Fishes**
   Robert J. Wootton
2. **Cichlid Fishes**
   Behaviour, ecology and evolution
   Edited by Miles A. Keenleyside
3. **Cyprinid Fishes**
   Systematics, biology and exploitation
   Edited by Ian J. Winfield and Joseph S. Nelson
4. **Early Life History of Fish**
   An energetics approach
   Ewa Kamler
5. **Fisheries Acoustics**
   David N. MacLennan and E.John Simmonds
6. **Fish Chemoreception**
   Edited by Toshiaki J. Hara
7. **Behaviour of Teleost Fishes**
   Second edition
   Edited by Tony J. Pitcher
8. **Genetics and Fish Breeding**
   Colin E. Purdom
9. **Fish Ecophysiology**
   J. Cliff Rankin and Frank B. Jensen

Forthcoming titles

Fish Swimming
J. Videler
Sea Bass
G. Pickett and M. Pawson
Fisheries Ecology
Second edition
Edited by T.J. Pitcher and P. Hart
Hake
Fisheries products and markets
J. Alheit and T.J. Pitcher
Impact of Species Change in the African Lakes
Edited by T.J. Pitcher
On the Dynamics of Exploited Fish Populations
R. Beverton and S. Holt
(Facsimile reprint)

Genetics and Fish Breeding

Colin E. Purdom

Ministry of Agriculture, Fisheries and Food
Fisheries Laboratory
Lowestoft, Suffolk, UK

CHAPMAN & HALL

London · Glasgow · New York · Tokyo · Melbourne · Madras

Published by Chapman & Hall, 2–6 Boundary Row, London SE1 8HN

Chapman & Hall, 2–6 Boundary Row, London SE1 8HN, UK

Blackie Academic & Professional, Wester Cleddens Road, Bishopbriggs, Glasgow G64 2NZ, UK

Van Nostrand Reinhold Inc., 115 5th Avenue, New York NY10003, USA

Chapman & Hall Japan, Thomson Publishing Japan, Hirakawacho Nemoto Building, 6F, 1-7-11 Hirakawa-cho, Chiyoda-ku, Tokyo 102, Japan

Chapman & Hall Australia, Thomas Nelson Australia, 102 Dodds Street, South Melbourne, Victoria 3205, Australia

Chapman & Hall India, R. Seshadri, 32 Second Main Road, CIT East, Madras 600 035, India

In memory of my father whose skill in breeding Yorkshire canaries for show purposes instilled in me an early love of genetics.

Contents

Preface   xi
Series foreword   xiii

1   The scope of applied fish genetics   1

    1.1   Introduction   1
    1.2   Fish Cultivation   2
    1.3   Ornamental fish breeding   8
    1.4   Fishing   8
    1.5   Habitat Destruction   10

2   Sexuality and reproduction   12

    2.1   Gonochorism   12
    2.2   Primary sexual characteristics   13
    2.3   Secondary sexual characteristics   15
    2.4   Modes of reproduction   16
    2.5   Egg production   16
    2.6   Production of spermatozoa   21
    2.7   Fertilization   22
    2.8   Cryopreservation   22

3   Mendelian inheritance: I. Colour   25

    3.1   The beginning   25
    3.2   Simple gene effects   26
    3.3   Complex effects of alleles   28
    3.4   Interaction of genes   30
    3.5   Summary of colour genetics   42

4 Mendelian inheritance: II. Characters other than colour   44

4.1 Introduction   44
4.2 Scale patterns of carp   44
4.3 Scale patterns in sticklebacks   47
4.4 Body shape   53
4.5 Fin shape   56
4.6 Electrophoretic variations   59
4.7 Miscellaneous characteristics   63

5 Quantitative genetics: I. Metrical characteristics   65

5.1 Introduction   65
5.2 Polygenic inheritance   66
5.3 Heritability   68
5.4 Growth rate   69
5.5 Stocks and strains   76
5.6 Spawning time   79
5.7 Food conversion efficiency   81
5.8 Disease resistance   82
5.9 Selective improvement in fancy fish   83

6 Quantitative genetics: II. Population studies   85

6.1 Introduction   85
6.2 Blood groups   86
6.3 Electrophoresis of soluble proteins   89
6.4 Inferences from population data   94
6.5 Molecular studies   98
6.6 Inbreeding   101
6.7 Application of electrophoresis   104
6.8 Summary   106

7 Chromosomes of fish   110

7.1 Introduction   110
7.2 Methods in cytology   111
7.3 Chromosome numbers and gross morphology   119
7.4 Banding polymorphism   127
7.5 Linkage and chromosome mapping   127

8   Sex determination   134

    8.1   Introduction   134
    8.2   Sex ratio   135
    8.3   Sex linkage   137
    8.4   Sex chromosome maps   141
    8.5   Sex reversal and sex determination   143
    8.6   Hybridization and sex determination   145
    8.7   Gynogenesis and sex determination   146
    8.8   Cytology of sex chromosomes   148
    8.9   An oddity, the H–Y antigen   156

9   Hybridization   158

    9.1   Introduction   158
    9.2   Hybridization in nature   159
    9.3   Artificial hybridization   166
    9.4   Summary of applied use of hybrids   174
    9.5   Hybridization and introgression   175

10 Atypical modes of sexuality   178

    10.1   Introduction   178
    10.2   Normal hermaphroditism   178
    10.3   Abnormal hermaphroditism   184
    10.4   Female-only species   185

11 Control of sex ratio   192

    11.1   Introduction   192
    11.2   Methods of sex ratio control   194
    11.3   Commercial application   201

12 Chromosome engineering   204

    12.1   Introduction   204
    12.2   Meiotic diploid gynogenesis   206
    12.3   Mitotic diploid gynogenesis   210
    12.4   Androgenesis   212
    12.5   Induced triploidy   213
    12.6   Induced tetraploidy   221

13   Gene manipulation   223

     13.1   Introduction   223
     13.2   Choosing the gene   224
     13.3   Inserting the gene   226
     13.4   Assessing the success of gene transfer   227
     13.5   Expression of the transferred genes   231
     13.6   Future developments in molecular genetics   232

14   Featured fish: I. Ornamental species   233

     14.1   Introduction   233
     14.2   The goldfish   233
     14.3   Koi – ornamental common carp   235
     14.4   The toothcarps   238
     14.5   Other species   239

15   Featured fish: II. Cultivated species   241

     15.1   Introduction   241
     15.2   Common carp   241
     15.3   Salmon and trouts   242
     15.4   Tilapias   244

References   246
Index   271

Preface

I hope that this book on fish genetics is complete in that it deals with genetics for all aspects of fish breeding, whether under captive conditions or within the natural environment. I have sought to bring a lifetime's interest in fish plus an almost equally long-standing interest in genetics together, into a framework which provides for a functional assessment of these two subject areas. The breeding of fish under artificial circumstances for food or for recreation or aesthetic needs is of world wide importance. The sustaining of natural populations is likewise of equal importance, whether in connection with the harvesting of the fish for food or in terms of the conservation of natural species against environmental harm. Although genetics is implicated in the consideration of such matters it should not be invoked wantonly, but used in a functionally practical way. It is hoped that this book will be of practical value to scholars and breeders and to anyone else with an interest in fish by presenting a balanced account of practical fish genetics. It is frequently the case that fishery managers - using the designation in its widest sense - are unfamiliar with genetic principles. Likewise geneticists often lack basic knowledge of fish. I hope I can partially rectify both limitations.

The book has a strong historical backbone for which I make no apology. Genetics has a relatively short history anyway, and its use as a framework for applied work has two important connotations. First, it may reduce the tendency in today's hectic research world to 're-invent the wheel'. Secondly, and of more importance, it may reduce the wasteful temptation now prevalent in grant-seeking activity, to reward the new research tool rather than the established approach irrespective of the use to which they may be put. It is the applied side of fish genetics that has given this subject its current great popularity - it should be encouraged further.

This book is not a detached, annotated bibliography but tries to tell a tale. For this reason, some aspects of basic genetics are repeated in more than one chapter. Genetics is not represented by a linear array of concepts but by a web of interdependent ideas. I hope that the book will be informative to those needing information or simply interested in fish genetics. I hope also that it will be stimulating to those already expert in the subjects covered but still open minded about them.

Many colleagues have helped in the production of this book consciously or not. My sincere thanks go to all of them, but in particular, to my secretary Jennifer Bugg for unending patience and word processor skills, to Irene Gooch and Martin Roach for graphics, especially the beautiful line drawings, and to Barbara Turner for making the extensive reference list comply with editorial standards.

C. E. Purdom

Series foreword

Among the fishes, a remarkably wide range of biological adaptions to diverse habitats has evolved. As well as living in the conventional habitats of lakes, ponds, rivers, rock pools and the open sea, fish have solved the problems of life in deserts, in the deep sea, in the cold antarctic, and in warm waters of high alkalinity or of low oxygen. Along with these adaptations, we find the most impressive specializations of morphology, physiology and behaviour. For example, we can marvel at the high-speed swimming of the marlins, sailfish and warm-blooded tunas, air-breathing catfish and lungfish, parental care in the mouth-brooding cichlids and viviparity in many sharks and toothcarps.

Moreover, fish are of considerable importance to the survival of the human species in the form of nutritious, delicious and diverse food. Rational exploitation and management of our global stocks of fishes must rely upon a detailed and precise insight of their biology.

The *Chapman and Hall Fish and Fisheries Series* aims to present timely volumes reviewing important aspects of fish biology. Most volumes will be of interest to research workers in biology, zoology, ecology and physiology but an additional aim is for the books to be accessible to a wide spectrum of non-specialist readers ranging from undergraduates and postgraduates to those with an interest in industrial and commercial aspects of fish and fisheries.

Genetics and Fish Breeding by Colin Purdom comprises the 8th volume in the series. This book is fascinating, rigorous and synoptic account of a lifetime working on fish genetics at the famous Lowestoft fisheries laboratory in UK.

The book commences with fundamental reviews of Mendelian inheritance, quantitative genetics, chromosomes, and sex determination in fish which will be useful to newcomers to fish work. Dr Purdom is no stranger to controversey and in this book he pointedly argues his case for lack of a genetic signal in the work on growth rates of fishes. From an applied perspective, he covers the topics of hybridization, and chromosome and gene manipulation in aquaculture. Sections focused on ornamental fish include material of interest to aquarists and anglers in addition to the more usual audience of fish ecologists and fisheries biologists.

Not only is an insight of genetics at molecular and population levels essential for an understanding of evolutionary change, but knowledge in this area is also important in assessing the sustainability of fish populations exploited by humans. This book aims to inform readers about fish genetics from both of these perspectives. In addition to meeting effectively these utilitarian aims, Colin Purdom's lucid text should endow readers with a sense of wonder.

Professor Tony J. Pitcher
Editor, *Chapman and Hall Fish and Fisheries Series*
Director, Fisheries Centre, University of British Columbia, Vancouver, Canada

Chapter one

The scope of applied fish genetics

1.1 INTRODUCTION

Two-thirds of our planet is covered by water and most of it supports fish life. This is enough of itself to make the study of fish, or any aspect of their lives, a compelling attraction, but several other reasons also exist. Fish, including the teleosts (bony fish) and elasmobranchs (sharks and rays), comprise almost as many species as all the other vertebrates put together; they span an enormous range of adaptive variation from the tiny parasite candiru, *Vandellia cirrhosa*, which normally occupies the gill cavity of larger catfish of the Amazon basin, but can make quite unpleasant alternative choices if human swimmers are available, to monstrous but benign travellers of the oceans like the basking shark, *Cetorhinus maximus*. They occupy a vast range of habitats with great environmental variety, spatial and temporal, and because of their poikilothermic physiology, their cold-bloodedness, they show an equally large range of adaptations. Fish occupy the aquatic environment; they show occasional short-term capabilities for terrestrial and aerial life but these are exceptional. The aquatic environment itself represents the earth's 'last wilderness' – this term is often used for Antarctica but it could more comprehensively be applied to the aquatic environment generally, which is much less well observed and therefore understood than the terrestrial environment. Man is making large impacts on the aquatic environment by diverse means and this is in danger of proceeding at a pace greater than the growth of our understanding of this important component of the environment.

Environmental issues are dominant in much of our political and economic thoughts and the aquatic environment is a vital, perhaps the vital, component of the world environment. Fish are a major part of this environment and exist at all levels from the close confinement of the desert pupfish – a simple hole in the Arizona desert – to the broad oceans of the world which support the wandering tuna fish. But, in addition to all of this, fish serve Man as a supply of very high quality food and as an unending source of recreational or aesthetic pleasure.

The study of fish genetics is relevant to all of these considerations. Direct implications are obvious for fish cultivation in all its many forms, but the genetic consequences of exploitation by commercial fishing must also be considered, and any habitat change must also have its genetic consequences – form and function are amalgams of the inherited potentialities of organisms and the environmental constraints applied to them, and change in either has a genetic consequence. Fish genetics is of relevance to fish breeding, whether in the context of fish cultivation or the exploitation of fish by fishing (Miller, 1957), or of the impact on fish of habitat change. This first chapter explores these three realms of fish breeding; later chapters cover all of the individual aspects of fish genetics and the extent to which they may be applied to or react to the pressures of domestication or exploitation.

1.2 FISH CULTIVATION

The farming of fish is reputed to be very old with a history going back thousands of years in China and several hundred in Europe. It is easier, however, to justify the view that it is Man's most recent exploration of the potential for domestication of a range of animal types. The older references to fish farming relate mainly to two freshwater species, closely allied to each other: carp, *Cyprinus carpio*, in the context of fish for food, and goldfish, *Carassius auratus*, for aesthetic purposes. Some references to early Roman/Greek maintenance of red mullet, *Mullus surmelatus*, in captivity are probably apocryphal, they are not easy fish to keep and feed and have not yet been bred in captivity. The weight of evidence strongly suggests that fish farming is a modern development deriving its strength in part from the overfishing difficulties experienced in commercial fishing, but also in part from the idealistic concept that Man should farm, not hunt. Over all these issues, the pragmatism of local production for local needs is probably the most productive of the forces which led to the successful development of fish farming for a wide range of species.

The species currently exploited in fish farming comprise several salmonids of freshwater and anadromous life styles, carp of Asian and European origin, catfish of the ictalurid family and eels, mullet and milkfish with wide salinity tolerance. Purely marine fish used for farming include the yellowtail tuna, *Seriola quinqueradiata*, in Japan, the turbot, *Scophthalmus maximus*, in Europe and a host of other flatfish species which modern fish farming technology has embraced in recent years. Production levels world-wide have increased steadily over the past decades as exemplified by the data from Japan (Fig. 1.1). They seem likely to continue to increase. The plethora of species is also likely to be expanded, but the farming practices used in their cultivation are now well defined, and it is these which are meaningful in terms of the genetic development of fish.

Some forms of fish farming depend upon the capture of 'seed' from natural resources. Eels, *Anguilla* spp, for example are farmed widely in the Far East and particularly in Japan, where production figures exceed 30 000 tonnes per year. The young elvers used to stock the farms are collected from the sea during their migration into fresh water, and there is therefore no possibility

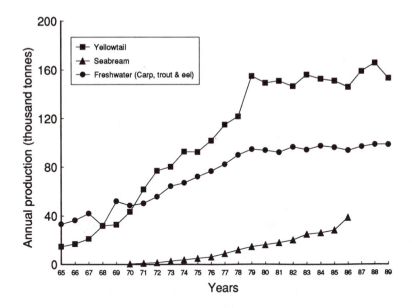

Fig. 1.1 Finfish production by aquaculture in Japan, 1981-1985 (OECD, 1989).

of genetic improvement. Research on the induced spawning of eels and hatchery techniques to generate seed have been successful, but the practicalities of hatchery production are so elusive that the chances of such technicalities competing successfully with the practice of natural harvesting seem remote. Other highly successful fish farming ventures which depend upon natural production of seed include yellowtail production in the inland sea of Japan and milkfish, *Chanos chanos*, mullet, *Mugil* spp., culture, widely practised in tropical regions of Asia. It is possible that all forms of fish farming originate from this basically simple practice of fattening fish in captivity, but it is a truism that genetic breeding improvements can only come about once breeding itself is the major way of providing seed.

Fish farming starting with controlled reproduction can be categorized under three headings:

1. intensive culture: uses high stocking densities in terms of fish weight per cubic metre of holding facility and has total dependence on artificial feeds;
2. extensive culture: employs low stocking densities with some reliance on natural productivity augmented by various levels of supplementary management;
3. ranching: uses some level of extensive or intensive culture to produce juvenile fish, which are then released into the natural environment where natural foods provide all sustenance required for growth of the fish to harvestable size; management after release is largely confined to simple habitat modification, such as reef building or predator control, plus harvesting.

One oddity perhaps needs detailing before we explore the nature of these fish culture categories more fully and the ways in which they may be subject to a genetic approach. This is the vallée system of eel production whereby elvers are collected from the natural environment and released into large inland lakes which otherwise would not receive a migratory run of these fish. Such practices support commercial fisheries in Italy and in Ireland, and are a special case without direct genetic implications per se but with possible genetic impact for other fish or fisheries in the areas where releases are made.

Intensive fish farming

In many respects this is the most artificial form of fish culture and subject to precise management control, and therefore amenable to sophisticated genetic management practices. The alternative view may also be taken that management options are so wide that genetic considerations are actually less needful than for the other forms of fish farming.

The family of fishes best known for intensive culture is the salmonids. Salmon and trout are farmed intensively throughout the temperate regions of the world and the basic technology has been in existence for over 100 years (Maitland, 1887) although much improved in modern practice (Sedgwick, 1990). Dramatic increases in production rates have been achieved in recent years (Fig. 1.2).

Artifice is the keyword in fully intensive trout farming. Natural reproduction is not permitted since it requires space, a profusion of flowing water, mixed gravel to several centimetres depth and an abundant patience by the cultivator

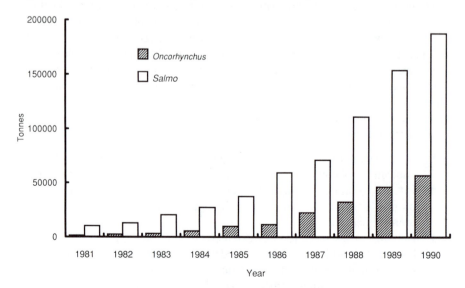

Fig. 1.2 World trends in salmon farming – the fastest growing sector of aquaculture (data from Seigel, 1990).

over the developmental sexual status of the stock. It is far simpler to await the phases of full sexual maturity of males and females – neither of which are short-lived phases – after which eggs and spermatozoa can be removed from individual potential parents and after scrutiny, either casual or objectively defined, mixed to create fertilized eggs. The eggs are washed clean of surplus milt, placed in flowing water in troughs or flasks and left to incubate without further management action other than the occasional treatment with anti-fungal agents, or the physical removal of the dead eggs which constitute the fungal threat.

On emergence from the egg, the young trout are still independent of their environment for nutrition and subsist on the contents of the yolk sac for a period of a few weeks. During this phase the young fish (alevins) are quiescent – under natural conditions protected beneath a layer of gravel – until the yolk reserves are almost fully exhausted, after which the fish fry swim up towards the surface of the water and begin to hunt for food. This is provided by the farmer as floating pellets of very small size to start with, increasing as the fish grow larger. The pellets contain protein, usually of fish origin, low levels of fats, added vitamins and binding material, together with a colouring supplement if pink-fleshed fish are sought. Antibiotics are included as required. This factory-made diet serves the fish from swim-up to slaughter or to sexual maturity, should the fish be chosen for future broodstock. It is wholly artificial and can be provided at various levels, within reasonable ranges depending on temperature, to control growth rate without undue physiological impact. Most fish are able to thrive on meagre or generous diets, and whether a large specimen in an abundant lowland environment is fitter than a small one in an upland 'health farm' is a largely philosophical question.

Intensive fish farming uses little space but very large volumes of water. These are required simply to provide sufficient oxygen for respiration – past attempts to remove this limitation by the injection of pure oxygen were biologically successful but, seemingly, economically non-viable. These very highly intensive approaches generated conditions where metabolites, princi-pally ammonia and nitrites, still did not limit production.

Intensive farming is thus highly artificial and not very dependent on the natural productivity of the environment. Temperature is the main environ-mental limitation but little can be done about it. In open-flow intensive farming systems, the most water-efficient use is about 5 million litres per day for every tonne of fish produced per year, and the economic prospect for heating or cooling this amount of water is very poor. Some attempts to reduce the need for water by recycling have been scientifically rewarding but not beneficial commercially. Light is another environmental variable with signif-icance for the physiology of fish. It is particularly important for sexual development (Bye, 1984) but may also influence feeding rate or food conver-sion and hence growth. It can be manipulated more or less routinely to extend control over specific events such as sexual maturation, but is not so easily applied for general purposes.

Salmon, trout, charr and other members of the family Salmonidae are farmed in this way. Less ubiquitously so are sundry marine fish, including turbot and sea bass, *Dicentrarchus labrax*. In these latter cases pumped seawater is usually needed and this implies a high energy cost. Less dependence on pumped supplies is achieved by a small modification of the intensive plan, the maintenance of fish in cages in open water where natural currents, tide- or wind-generated, promote water exchange. Cage culture is usually used for the later stages of on-growing, not for small fry for which mesh sizes would present management problems, particularly through clogging due to the growth of settling organisms. The most notable expansion in recent years in fish cultivation using cages has been for Atlantic salmon, *Salmo salar*, in the Northern Hemisphere, but equally large scale operations have been undertaken for many years past in Japan for the culture of yellowtail tuna.

Extensive fish farming

The definitive difference between this and intensive farming is, of course, space – intensive systems are rated at kilograms of fish per cubic metre whereas extensive systems are in terms of tonnes per hectare, but the differences are more fundamental than this. The principal feature of extensive farming is that much of the production is based on natural foods but augmented by additional feeds, often of a poorer protein content than those used in intensive culture. The reason for this is that the natural productivity of zooplankton and benthic organisms represents very high quality animal protein. The key management role in extensive farming is therefore towards creating a suitable environment for natural productivity.

The most well established extensive farming practice is that for the common carp, *Cyprinus carpio*. This species is farmed all over the world where appropriately warm temperatures for good growth are found, i.e. 20-30°C, but what follows is the basic practice in Eastern Europe and the CIS.

Carp farming concentrates on natural as opposed to artificial procedures. Spawning is natural; selected parents, often one large female and two or more males, are placed in small (0.1 ha) shallow ponds where they spawn on natural vegetation. Following spawning, the parents are removed, the eggs hatch and the free-floating fry soon begin feeding on natural zooplankton which is encouraged by control of water quality factors, such as pH, and nitrate and phosphate levels. This water quality management is usually started well before spawning so that natural cycles of plankton production can get under way in time for the emergence of the fry.

When young fish reach a weight of 25 g or so - possibly only after one full season of growth – they are transferred to larger ponds (1 ha or more) where they may spend a further season of growth to produce fish of 500 g weight or larger. The on-growing ponds are purpose built so that they can be emptied of water for harvesting and for water hygiene. Water quality is controlled as in the spawning ponds, and supplementary diets are given, often made from cheap waste materials such as vegetables and brewery or abattoir waste. As a rule of thumb measure, natural productivity can be one or two hundred kilograms per hectare per year; following water quality control and fertilization this can reach 500-600 kg ha^{-1} year^{-1}, but with supplementary feeding as much as 5 tonnes ha^{-1} year^{-1} can be achieved - thus although carp farming exploits natural circumstances, the real expansion only comes with the provision of artificial diets.

Common carp is the most widely farmed species of fish, but very significant production of channel catfish, *Ictalurus punctatus*, takes place in the southern part of the United States (Tucker and Robinson, 1990) using similar if somewhat more modern methods than those that are used in carp farming in Eastern Europe. A wide variety of fish of the families Cyprinidae and Siluridae is also farmed in Asia. A further group of fishes farmed extensively includes the tilapias, e.g. *Oreochromis* spp. No truly marine fish are farmed in this way, presumably because of the lack of adequate control of the productivity of marine ecosystems. Brackish water practice is feasible, however, for such species as mullet, *Mugil* spp., and milkfish, *Chanos chanos*, but these activities still largely rely on natural seed and are therefore beyond the scope of genetic improvement.

Sea ranching

Sometimes also described as stock enhancement, the procedure of rearing fish to a size suitable for release into the sea, or indeed into any large body of water, to complete growth to marketable size before recapture is one of the more forward-looking aspects of fish farming. Its greatest appeal is that is can represent net production of fish. Other forms of fish farming are net users of protein resources. In intensive farming it may take three units or more of feed protein to make one unit of fish protein, and extensive culture is only significantly better than this if production levels are low and mostly provided for by natural production. Sea ranching exploits the natural environment for the major part of the growth phase. A further point is that most of the present forms of sea ranching use migratory species which return to the point of release – so no capital expenditure in the form of ships is needed, nor energy in the form of fuel, for the harvesting of the crop (Joyner, 1976).

Throughout the world salmon are by far the most popular species for ranching, but significant developments have also taken place for sturgeon, another group of anadromous fish which spend most of their growth phase in the sea whilst spawning in fresh water. Marine fish species are also being exploited in this way, and once again Japan, that most adventurous of fish farming societies, is taking the lead.

The two anadromous fishes, salmon and sturgeon, neatly illustrate the intensive and extensive components of the practice of sea ranching. Young salmon are reared to the size at which they go to sea in exactly the same way that trout are reared intensively, whilst sturgeon fry inhabit more or less natural lakes, feeding on natural plankton before being ushered into their downstream journey by the draining of the lake in which they have lived.

Recapture of salmon and sturgeon by trap or net on their return migration serves both for harvest and for provision of spawners for the next round of rearing. It is this closed cycle of production which makes a genetic approach to breeding and improvement feasible in ranching.

1.3 ORNAMENTAL FISH BREEDING

Fish cultivation is not just for food production. The breeding of fish of all sorts for the sheer technical satisfaction of doing it, or, more commonly, in order to appreciate the aesthetic beauty these animals display, is probably just as important as fish farming for food. Breeding such fish for sale is of very great importance world-wide as a trade and as a hobby.

The breeding of fancy fish is conducted professionally on large fish farms and also on a very small scale by individuals who rate it as a hobby. It probably involves far more practitioners than other forms of fish farming and generates highly significant financial turnover in rich and poor nations alike. Very many different species of fish are reared, including freshwater and saltwater forms, but they are mostly warm-water lovers and, in particular, include carps of the families Cyprinidae and Cyprinodontidae, and exotic species of the family Cichlidae. The cold water salmonid or esocid (pike) families are not favoured. Breeding methods are diverse and tailored to the needs of individual species, but one thing they all seem to have in common is that years of effort have generated distinctive genetic varieties in most of the species bred.

1.4 FISHING

Commercial capture of fish by trap, net or hook and line might almost be equated to the primitive hunting of mammals and birds which our Iron Age ancestors pursued to provide themselves with food. Many people have likened the expansion of fish farming to the equivalent of the social advance that saw off hunting in favour of agriculture, but the situation is not quite the same: our conquest of the aquatic environment can never equal that of its terrestrial counterpart, and natural fish populations will always remain as vital sources of marine food, short of cataclysmic events which would have even more dramatic consequences elsewhere. This is not to say that fish populations remain static – they most certainly do not, and in most circumstances decline sharply once exposed to exploitation as a major fishery (Cushing, 1988). The total world catch of fish seems still on an upward trend (Fig.1.3), but this embraces a widening of the effort plus the expansion of some fisheries and the near collapse of others. At its present total of 99 534 600 tonnes for all species of fish and shellfish in 1989 it is believed by some to be at its maximum level (FAO, 1991).

Figure 1.4 shows the changing fortunes of North Sea cod, *Gadus morhua*, and sole, *Solea solea* – two quite different fish which show the consequences

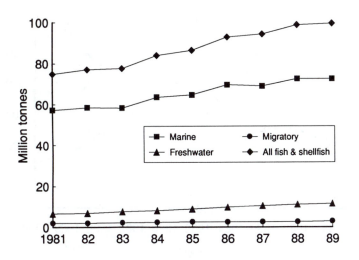

Fig. 1.3 World fish capture, including aquaculture, 1980-1989 (FAO, 1991).

of overexploitation: a long slow decline for cod whilst sole landings have collapsed dramatically from their high point of 3500 tonnes in 1963 but now continue at a lower sustained level. For fish farming, the significant feature of Fig. 1.4 is that the unit value of fish increases with diminishing landings. Value is all important in fish farming. However, it is not the pattern of catch statistics which will have genetic consequences but the manner of fishing and particularly the extent to which fishing exerts a selective pressure.

One of the most obvious of selective measures in fishing is the size limit widely used to avoid wasteful destruction of fish before a sensible growth yield, or capture before at least one opportunity to spawn. Even in the latter case, however, the spawning structures of exploited populations still remain se-verely curtailed. A further constraint is the application of closed seasons, which are often used to attempt to protect fish during spawning. This is a common feature of salmonid fisheries, but, since migration and spawning are not accurately matched, a close season for fishing may exert considerable influence on spawning time itself.

Finally, of the major constraints, the potential of some fishing methods such as purse seining can lead to the capture of entire populations, and the question then is to what extent the 'hole' will be filled by other populations with differing genetic structure.

These and other considerations of the genetic consequences of major fishery activities are just beginning to exercise the minds of conservationists. They need to be considered in a fish genetics context and not in the more popular, but sometimes misleading, terms of the ecologist.

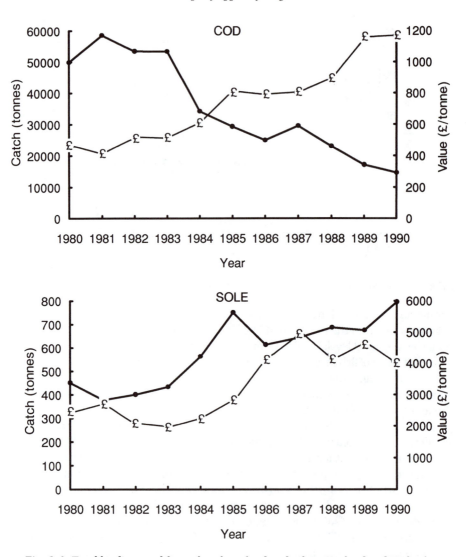

Fig. 1.4 Total landings and first sale value of cod and sole in England and Wales from the North Sea, 1980-1990.

1.5 HABITAT DESTRUCTION

Concern is expressed today at all forms of environmental perturbation generated by Man. Various levels of damage can be envisaged, from contamination, the lowest level of effect, defined as the presence of detectable man-made substances in the environment, through pollution – the level of contamination at which harm is discernible – to complete destruction of the environment itself. In its 1990 report on the aquatic environment, the United Nations Environment Programme Group of Experts on Scientific Aspects of Marine Pollution (GESAMP) found contamination – Man's imprint – in all of the world's oceans but significant pollution, and probably a worsening problem, only in coastal seas (GESAMP 1990). In inland waters the damage is even more apparent, whether in terms of simple oxygen deficiency in rivers below industrial complexes or in the form of major catastrophes such as that of the Aral Sea, which has virtually disappeared through over-use of irrigation. These are major examples of habitat destruction, but modification in inland waters is almost universal and, as stated at the outset of this introductory chapter, environmental change is as relevant as genetic change per se to the genetic status of populations of animals and plants - they do not live independently of the environment.

Chapter two

Sexuality and reproduction

2.1 GONOCHORISM

Sexuality is not synonymous with reproduction. There are many forms of reproduction in which sex plays no part, from the simple binary fission of unicellular organisms, bacteria to protozoa, to the complexity of parthenogenesis in multicellular organisms where individuals recreate themselves in perfect copy. But sexuality is the means whereby the genetic diversity of living things is expressed in terms of inherited patterns and evolutionary progress. These ideas will be developed more fully later, but sexuality itself, and the various modes of reproduction in fish, will be described because they are crucial to any understanding of fish genetics and to the application of the laws of heredity to fisheries development or protection.

The basic pattern throughout the animal and plant world is of sexuality involving separate male and female individuals: this is sometimes called **gonochorism**. A not uncommon form of sexuality in fish is **hermaphroditism**, in which males and females exist but individuals may be first one sex then the other or both male and female at the same time. This will be explored in greater detail later in this chapter along with other types of sexuality which might be termed atypical.

Gonochorism is the basic mode of sexuality and the majority of fish species express it, but in a variety of form and function which is quite exceptional within the vertebrates and exceeded only, perhaps, within insects in the animal kingdom as a whole.

The primary feature of the distinction between the sexes is clearly the nature of the gonad, whether testis or ovary, but in addition, many secondary characteristics exist which may be required to permit the mechanics of sexual reproduction, such as germ cell ducts and intromittent organs, or to promote sexual awareness or activity. For the latter, colour pattern, body shape and size, and even organs for the production of pheromomes and the reception of their messages are of importance. Table 2.1 lists those features which differentiate the sexes of fish, together with some examples.

Table 2.1 Features which differentiate the sexes in fish

Type of characteristic	Example
Primary characteristics	Gonads
	Hormones
Secondary characteristics	
(a) Requisite	Ducts,intromittent organs and ovipositors
(b) Accessory	Size, colour, finnage and many other aspects
(i) permanent	Colour in *Poecilia* and *Coris*
(ii) temporary	Colour and kype in *Salmo* and *Oncorhynchus*

2.2 PRIMARY SEXUAL CHARACTERISTICS

Figure 2.1 illustrates the basic form of the gonad and the requisite sex ducts of fish. Ovary and testis are normally paired organs lying within the coelomic cavity suspended by a dorsal mesentery which also supports the ducts for the ovary (Mullerian) or testis (Wolffian), which open to the outside via a genital pore. A number of minor variants exist, particularly in the association of the gonad and its ducts with the kidney and its ducts, but the only important example is where the ovarian ducts are incomplete and allow eggs to gather freely in the coelom – this is the situation in salmonids and some other fish families but the majority of fish species have complete oviducts which carry eggs directly to the outside.

The germ cells in all animals seem to be designated as such and set aside at an early phase in the developing embryo. They remain more or less quiescent until the somatic elements of the gonad, the external membranes and interstitial tissue, develop late in general organogenesis. In fish, this often takes place well after the fry has hatched and begun feeding. During organogenesis the putative germ cells migrate into position to form part of the gonad but only start to develop into oogonia or spermatogonia when the sexual differentiation into male or female form has taken place. Concomitant with this change, or possibly sequentially, the interstitial cells of the gonad (non-germinal tissue) begin to produce sex hormones which themselves determine

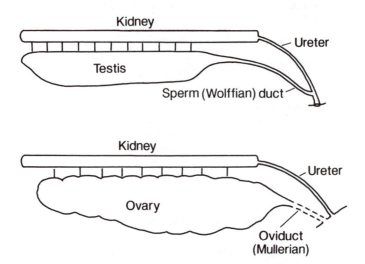

Fig. 2.1 The generalized layout of teleost gonads

all subsequent, secondary, sexual characteristics from dimorphism to sexual behaviour.

It is beyond the scope of this book to detail the hormonal control of sexual development and reproduction but a comprehensive treatment of this subject is given by Hoar *et al.* (1983). In general, however, the interstitial tissues of the gonads generate sex steroids in response to hormonal messages from the pituitary gland at the base of the brain. This pituitary-gonadal axis controls the expression of sexuality, including the development, maturation and release of gametes in response to climate or seasonal cues. By these means the sexual cycles of fish are closely tuned to environmental circumstances. It is often of advantage in practical situations to be able to overcome the limitations of these natural constraints on reproduction, and means to achieve this are possible by environmental manipulation of light or temperature, or by direct administration of hormones (De Vlaming, 1972; Donaldson and Hunter, 1983).

The nature of sex determination and its genetic control is described fully in Chapter 8, but for the present it is sufficient to point out that there is a lack of distinction between male and female primordial germ cells in teleost fishes (Reinboth, 1970) and to highlight the continuing uncertainty as to whether or not sex hormone production within the undifferentiated gonad leads to differentiation and to what extent the presumptive germ cells influence events. Whatever developmental processes are involved, the end result is remarkably well channelled with maleness and femaleness, being the overwhelming norms, and atypical deviations, such as pathological hermaphroditism, being extremely rare. It will become evident later that sex determination in fish can be regarded as plastic (for a review see Francis, 1992) but possibly not primitive. In the teleosts a great amount of evolutionary divergence is apparent within the general field of sexuality.

2.3 SECONDARY SEXUAL CHARACTERISTICS

Table 2.1 lists requisite and accessory secondary sexual characteristics. The former include the non-germinal tissues of the gonads and the gonadal ducts, which embryologically precede sex differentiation, and an array of features such as claspers and intromittent organs which develop after sex differentiation. These ducts serve primarily to conduct the gametes to the outside of the fish but, in addition, they probably generate components of the semen in the male and ovarian fluid in the female. This aspect of sexual physiology is not fully understood but has relevance to production, use and storage of gametes. A further function of the accessory glands on the ducts in some fish is to generate pheromones, the chemical messengers which can control much of the sexual behaviour of animals.

All of the accessory secondary sexual features, i.e. those that relate to courtship and other aspects of mating, develop after sex differentiation and can be permanent or temporary attributes of fish. In the latter case the sexual characteristics develop before reproduction and disappear afterwards. All of the secondary sexual characteristics seem to be controlled by steroid hormones produced in the gonads. Males typically develop more extreme characteristics of shape, colour and aggressive behaviour, and all of these traits respond to the production of androgens by the testis and can be influenced by the artificial administration of sex hormones such as the manufactured steroid methyltestosterone. This also applies to permanent features, such as the colour patterns shown by males of some cyprinodontids (Grobstein, 1948), as well as to temporary colour changes which are very typical of maturing salmon and sundry other fish species.

Female accessory secondary sexual characteristics are less obvious than those of their male counterparts and this has led to a view that femaleness is the undifferentiated or primitive state and maleness the derivative of it. The most obvious distinguishing feature of females across a wide range of species is their larger size. This is true of marine flatfish, many cyprinids such as the common carp, eels and sturgeons. Other groups of fish, however, show no sexual dimorphism for size – this is particularly common in shoaling or free swimming species such as the mackerels and salmonids. A few groups of fish, such as the cichlids, show male predominance in size, often in conjunction with positive sexual roles such as nest building and egg care.

One final feature of growth rate and sex is that even if the basic pattern reflects equal growth rates of the sexes, precocious or early maturity in males is often associated with cessation of growth leading to marked sexual dimorphism for size. The classic example here is the so-called precocious maturity in male Atlantic salmon (Thorpe, 1985) which frequently involves a small but variable proportion of the males in nature. Similar but more systematic variation may occur in brown trout, *Salmo trutta*, where all females in a population may go to sea and grow large whilst males remain in their natal streams and remain small, probably reflecting the poorer dietary opportunities in the river as compared with the sea.

These secondary characteristics are of very significant economic importance. The brilliance of colour patterns and elegance of fin extensions are of obvious value to the breeders of decorative fish whilst growth rate is of considerable interest to fish farmers. Colour and the behavioural traits of aggressiveness are also of importance for farmed species because of their impact on the appearance and hence the sale value of the fish. For all of these reasons, methods for the control of sex ratio and of sexuality itself are important in fish genetics. They will be described in Chapters 11 and 12.

2.4 MODES OF REPRODUCTION

These range widely, from the more or less simple synchronous release of eggs and sperm in shoaling fish such as herring, *Clupea harengus*, to specific pair mating with quite marked behavioural complexes as in most fish species. Advanced levels of parental care may take the form of nest guarding, as in the tilapias and sticklebacks, *Gasterosteus* spp., internal incubation of fry, as in *Sebastes* species and seahorses, *Hippocampus* spp., to internal nourishment as in the extraordinary cyprinodont *Ameca splendens* whose internally incubated fry develop extensions of the gills for food absorption 'in utero'.

For practical purposes simple egg and sperm release is the most convenient mode of reproduction for application in fish farming. The artificial admixture of eggs and sperm, bypassing natural procedures, is the most manageable system and allows for a variety of genetic techniques, such as chromosome manipulation and gene transfer, to be applied to breeding. It would not be practicable to apply these under natural conditions of reproduction.

2.5 EGG PRODUCTION

As explained earlier in this chapter, the primordial germ cells are set aside early in embryogenesis and migrate in late organogenesis into position to be part of the enveloping gonad. In the female, these undifferentiated stem cells then undergo cell divisions (**mitosis**) to produce populations of oogonia within the germinal epithelium of the ovary. These oogonia are also stem cells and divide mitotically to produce oocytes which no longer divide but eventually mature into eggs (Fig. 2.2). It seems probable that the primordial germ cells, which can differentiate into male or female germinal tissue, are replaced early in gonad formation by their differentiated derivatives which can then only produce either male or female gametes, not both. Nevertheless, the oogonia remain sufficiently undifferentiated to continue as stem cells to provide nests

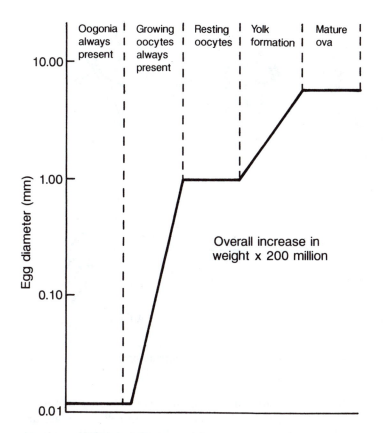

Fig. 2.2 Sequences in egg formation and growth in salmonid fishes.

of primary oocytes which persist throughout the lifetime of the female fish. This contrasts sharply with the situation in higher vertebrates where the mitotic activity of oogonia stops before sexual maturity, to leave a finite stock of developing oocytes.

Oogonia are typically small round undifferentiated cells having a large nucleus and very little cytoplasm. The first primary oocytes are similar to oogonia but, of course, do not divide any further: they enter a growth phase in which cytoplasmic content increases rapidly (Fig. 2.2). Nests of small primary oocytes can be observed very early in the life of an individual and long before the onset of sexual maturity in many species such as the marine flatfish, *Pleuronectes platessa*. In salmonids, primary oocytes are found in the ovary of the fry a few months after hatching even though sexual maturity will not occur for another 3 years or more. In other species, e.g. *Cyprinus carpio*, the gonad appears empty of developing germ cells for at least 1 year after hatching – this makes it very difficult to determine the sex of young carp, at least in temperate regions; in the tropics they can be fully mature within 1 year.

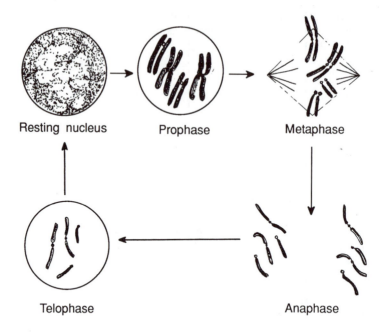

Fig. 2.3 Simplified diagrammatic representation of chromosome behaviour in mitosis - two pairs of chromosomes are shown.

Primary oocytes always seem to be present in fish ovaries once maturity is reached and it seems likely, therefore, that these germ cell stages are at a stationary or equilibrium phase of development. Further maturation involves two developments: first, a very considerable increase in size, and secondly, a sequence in which the genetic material in the form of **chromosomes** is modified by the reduction of chromosome number by half. This is the **meiotic cell division**, and the halving of chromosome number prevents, of course, any increase of genetic material which would otherwise occur each time an egg and a spermatozoan combined. The meiotic cycle also has other genetic consequences which will be covered later.

The size of eggs increases from a few micrometres in diameter up to over a centimetre in some species of fish although the normal range of maximum size in eggs for most fish species is 1–2 mm. The increase in size primarily comprises a huge intake by the oocyte of a glycolipoprotein called vitellogenin. This material is synthesized in the liver under the control of gonadal steroids and is incorporated into the developing egg as it lies within the ovarian follicle. At this stage the eggs are called secondary oocytes and they can appear to be very variable in size. As the time of reproduction approaches, the oocytes that will complete their development and become the definitive eggs mature to a very uniform size.

Egg size is important in fish culture. Large eggs generate large fry, and the problems of early feeding are less severe the larger the fry.

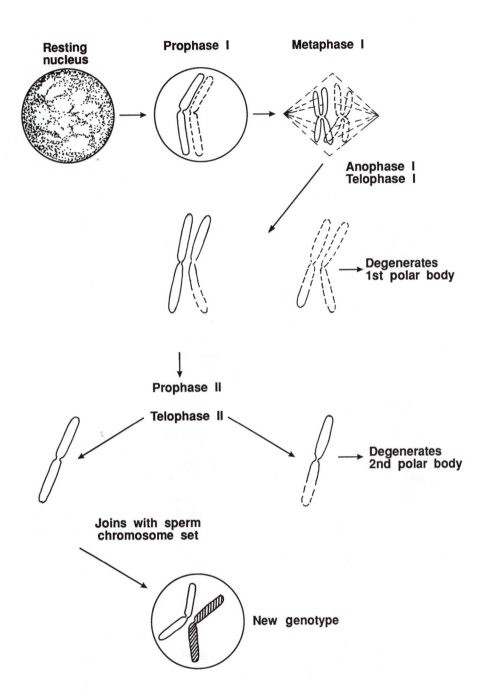

Resting nucleus

Prophase I

Metaphase I

Anophase I Telophase I

Degenerates 1st polar body

Prophase II

Telophase II

Degenerates 2nd polar body

Joins with sperm chromosome set

New genotype

Fig. 2.4 Chromosome behaviour during meiosis - representation of the fate of one chromosome pair, maternal in solid line, paternal dotted.

The real distinction between primary and secondary oocytes lies in the behaviour of the chromosomes, the microscopic constituents of the cell nucleus which contain the genetic material. The chromosomal content of the primary oocyte is the same as that of the mitotically active oogonium. Figure 2.3 illustrates the basic mechanics of chromosomes during mitosis and identifies the various phases of the cycle from prophase through metaphase and anaphase to telophase after which the nucleus is reformed. The chromosomes themselves are thread-like structures which contract at the start of mitosis to form rods, with one distinctive constriction called the centromere which acts as a locomotory organelle (further details of chromosome structure and terminology may be found in Chapter 7). The simple replication of chromosomes during prophase is accompanied by disappearance of the nuclear membrane and followed by segregation of the daughter chromosomes (chromatids) to opposite poles of the metaphase spindle by mutual repulsion of the new centromeres. The spindle arises in metaphase and appears in the form of two cones of fibres with bases opposed. This segregation in metaphase/anaphase leads to the formation of two new cells and guarantees that each daughter cell has the same chromosome complement as the original mother cell. Thus the primary oocyte has the same chromosome complement as any other somatic cell of the body, comprising a double set of chromosomes one of paternal origin, the other maternal. In the formation of secondary oocytes this diploid number of chromosomes is reduced to a single set (haploid) by the process of the meiotic cycle although it should be noted that the cycle is not completed until after the egg has been fertilized.

Meiosis comprises two separate cycles of nuclear division without an intervening cell division. Most of the important action takes place in the prophase of meiosis I (Fig. 2.4) before the chromosomes are microscopically visible. The maternal and paternal chromosome homologues first pair closely together to produce a haploid number of paired units or bivalents. Each chromosome is then replicated thus generating four-stranded entities still termed bivalents but comprising two pairs of **chromatids**. The replication may be visualized as starting at one end of the chromosome and proceeding along its length. As the replication proceeds, it may happen that the new thread on the maternal chromosome links across to the replicating paternal chromosome and vice versa. This is the basis of crossing over, which will be covered more fully in Chapter 7, by which sections of maternal and paternal chromatids are exchanged. At this stage the pair of chromatids is still described as a chromosome.

Centromere replication does not occur during meiosis I and thus the mutual repulsion forces during metaphase and anaphase separate the maternal and paternal centromeres within each bivalent. The paired chromatids aggregate at the poles. Only a single pair is shown in Fig. 2.4 but in reality many pairs exist. Centromeres of maternal and paternal origin, respectively, do not move together but behave independently such that the new groups of paired chromatids comprise a mixture of original maternal and paternal chromosomal material. One of the groups degenerates – it is called the first polar body. The other forms the putative 'nucleus' of the secondary oocyte, although nuclear membranes do not form and the chromosomes begin the second phase of the meiotic cycle.

The oocyte 'nucleus' in the secondary oocyte still contains the diploid quantity of chromatids, but only the haploid number of centromeres. In the next phase, the second part of meiosis, the centromeres divide and a second metaphase is created in which the chromatids become chromosomes and again segregate to opposite poles. Once again one lot degenerates (the second polar body) whilst the other persists as the female pro-nucleus containing a haploid set of chromosomes. This second phase of meiosis is delayed in the maturing egg, however, and is completed only after fertilization. At this suspended phase, the secondary oocyte now undergoes its considerable growth to attain the final egg size.

The secondary oocyte thus far is embedded in the tissues of the ovarian epithelium and surrounded by theca cells which provide for the transfer of vitellogenin and other materials for incorporation into yolk. After full growth, and subject to hormonal control already touched on, the secondary oocyte becomes an egg, and is liberated into the lumen of the ovary or into the coelom, the body cavity. This process of ovulation may be a long drawn out event taking several days, as in salmonids where the eggs lie free in the coelom, or quick, as in flatfish where successive batches of eggs are ovulated each night over a period of weeks. In the latter case, the eggs have only a short life and become 'over ripe' if not fertilized within a few hours of ovulation. Salmonid eggs may be retained in the fish for 10 days or more. Egg quality, an oft used but rarely defined characteristic in fish farming, probably depends as much on post-ovulatory age as on any other factor or combination of factors.

As soon as fertilization occurs, the second metaphase spindle forms and the female meiotic cycle is completed with the ejection of the second polar body and the subsequent fusion of the chromosome complement of the sperm cell with that of the egg cell to re-establish the diploid state – mitotic divisions then ensue, followed by embryonic and organogenic processes leading to the formation of the new individual.

2.6 PRODUCTION OF SPERMATOZOA

The process of meiosis in the male is similar to that in the female except that it is a continuous process leading from diploid spermatogonia to haploid spermatozoa. Polar bodies are not formed in spermatogenesis – all of the products of meiosis end up as spermatozoa.

Production of spermatozoa in fish is normally continuous over the spawning period, and although less is known about its dynamics than is known about ovulation, it seems probable that the quality of the spermatozoa, i.e. their ability to successfully fertilize eggs, does decline with time

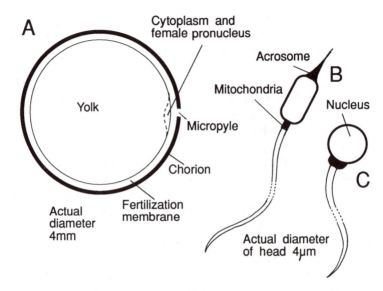

Fig. 2.5 Basic structures in (A) the fish egg and in spermatozoa from(B) sturgeon and (C) salmon.

after release of the mature cell into the lumen of the testis. On release into water or ovarian fluid the swimming activity of the spermatozoan increases rapidly but only for a very short time, after which the cells become immobile.

2.7 FERTILIZATION

The structures of fish eggs and spermatozoa are illustrated in Fig. 2.5. The critical features for later genetic studies are the micropyle and the membranes of the egg, and the acrosomal region of the sperm head which effects entry through the egg cell membranes in some species.

Fertilization takes place via the micropyle, a narrow channel through the outer egg membrane. It is normally accomplished by the spermatozoan swimming to the egg surface, locating the entry to the micropyle and swimming down it to make contact with and penetrate the inner egg membranes. The fertilization membrane lifts after penetration by the first spermatozoan to reach it, and this is probably the basis of the mechanism for avoidance of polyspermy (Ginsburg, 1968) which, if permitted, would have adverse genetic consequences.

2.8 CRYOPRESERVATION

The storage of gonads or even of embryos has long been advocated for a variety of applications such as conservation of genetic resources, either in the form of domesticated stocks or of natural, perhaps endangered species, or as a tool

Table 2.2 Cryoprotective diluent for fish spermatozoa

125 mmol l^{-1}	Sucrose
100 mmol l^{-1}	Potassium bicarbonate
6.5 mmol l^{-1}	Reduced glutathione
10% (v/v)	Dimethyl sulphoxide
5 – 20% (v/v)	Egg yolk

in normal breeding for routine or special use. Indeed, one of the first successful uses of frozen spermatazoa in fish breeding was to enable cross-fertilization of spring and autumn spawning herring (Blaxter, 1953).

The freezing of herring sperm was achieved with solid carbon dioxide at a temperature of –40°C. Following the successful use of deep-freeze storage of semen in artificial insemination (AI) practices with cattle, several studies with fish made use of the much lower temperature of liquid nitrogen (–120°C). Successful long term storage of spermatozoa has been achieved by Mounib *et al.* (1968) for cod, *Gadus morhua*, by Pullin (1972) for plaice, *Pleuronectes platessa*, by Chao *et al.* (1975) for grey mullet, *Mugil cephalus*, and tilapia, and by Cognie *et al.* (1989) for carp, *Cyprinus carpio*. Several authors have reported on the cryopreservation of rainbow trout, *Oncorhynchus mykiss*, milt, including Billard (1977), Stoss *et al.* (1978), Baynes and Scott (1982) and Scott and Baynes (1980) who reviewed the then available literature for salmonids. More recent reviews of cryopreservation and physiology of gametes in fish are by Stoss (1983) and Leung and Jamieson (1991).

Other than temperature of storage, important aspects of cryopreservation of milt are the nature of the diluent and its cryoprotective component, the rates of cooling and of thawing prior to use and the viability of the spermatozoa at the start.

The cryoprotective diluents mostly follow mammalian practice using ethylene glycol, glycerol or dimethyl sulphoxide together with more exotic items such as phosphatidylcholine (lecithin) or egg yolk. The purpose of the diluent is to protect the cell membranes from physical damage by the passage through them of ice crystals as they form. Baynes and Scott (1987) describe a modification (Table 2.2) of Mounib's original recipe (1978) for a cryopreservation diluent in fish.

In trials with rainbow trout milt, the success with which cryopreserved spermatozoa were able to fertilize eggs was significantly improved by the addition of egg yolk, but the magnitude of the effect was compounded by variations between batches of milt from different males. The recipe with 10% egg yolk gave the most consistently high results, but despite this the technique was only recommended for small batches of eggs and not for routine use. This seems to be the major theme in attempts to cryopreserve milt – much variation between males and low expectation of wholesale success. The most important problem probably resides in the physiology of the maturation of spermatozoa within the seminal fluid generated by the sperm ducts.

Cryopreservation of eggs has been attempted without much success so far

(Stoss, 1983). The major problem must be the complex structure of the egg, particularly within the egg membranes and the yolk/cytoplasm alignment. Small, non-yolky eggs of the type produced by mammals and molluscs do seem amenable to freezing and thawing.

Despite the limitations of present practice with the deep freezing of gametes and fertilized eggs, the obvious benefits should encourage further research. Meanwhile, other less comprehensive techniques should be explored. The extension of the life of gametes by storage at low temperature is an obvious and practical solution to short-term problems. Current practice with AI in pigs uses diluted chilled semen and this approach seems well suited to modern methods of trout and salmon farming. Delayed hatching of eggs by holding them at low temperatures is also possible so as to extend current spawning seasons. A more radical approach might be possible if the physiology of diapause in certain genera of the killifish family (Cyprinodontidae) was more fully understood. The eggs of these so-called 'annual' fishes can survive in suspended development for many months – an adaptation to life in temporary water bodies. To maintain such suspended animation generally in fish embryos would be enormously beneficial.

Chapter three

Mendelian inheritance: I. Colour

3.1 THE BEGINNING

The fact of heredity, that like begets like, has probably been obvious to Man since he first began to keep animals and then domesticate them, some 12 000 to 14 000 years ago. Mechanisms of inheritance must also have exercised his mind from much the same time because our intelligence and curiosity cannot have changed much over such a short evolutionary period, but the flowering of this interest really took place very recently, during the 19th Century, throwing up such concepts as Lamarck's theory of the inheritance of acquired characters and Darwin's hypothesis of pangenesis – that the germ cells contained a representative of all the cellular components of the fully developed body! Gregor Mendel brought rationality to the confusion through careful simple experiment and showed that inherited material came from both parents in the form of discrete units which did not blend in the offspring but were passed on to the next generation in the form they had been received from the previous one. Thus the concept of the 'gene' was established and the study of inheritance became the science of genetics. Subsequent study of visible components of cells generated the view that the genes were carried on chromosomes, with many more than one gene per chromosome and, into modern times, the chemical nature of the genetic material was defined as deoxyribose nucleic acid (DNA) with a chemical structure involving purine and pyrimidine molecules polymerized with sugar moieties and phosphate bonds into a complex macromolecule. The Watson–Crick proposal for a double helical structure involving the four base molecules in complementary union to provide purine–pyrimidine pairs suggested a mechanism for the DNA coding of the molecule and also the means whereby it replicated. This laid the foundations for the establishment of the nature of the genetic code and the beginnings of molecular biology. This amazing transformation, from the barely conscious views of pre-Mendelian scientists to the seemingly effortless ease by which modern geneticists can now synthesize genes themselves, took little more than a century and made little progress in fact until the second 50 years. Molecular genetics itself owes little to fish research but it is now having an impact on applied fish biology by the production of genetically modified fish (Chapter 13).

The existence of genes as discrete units which determine characteristics of living things, their presence in alternate forms or **alleles**, and combination in pairs in individuals is the simple binary code of genetics. Like the binary code in computers, it is simple taken step by step but capable of quite enormous complexity when looked at holistically. As with computers, the simple binary code makes genetics a very methodical and a very precise science – at least in theory.

The simple expression of alleles in inheritance is often described as Mendelian. Let us look at some examples of Mendelian inheritance in fish and explore the complexities which can arise when more than a single step is assessed and some of the consequences of these for fish breeders.

Simple Mendelian inheritance is the backbone of genetics. Studies in fish in this context began soon after the rediscovery of Mendel's published work in 1900, some 35 years after it had first appeared in a little-known scientific journal. Continuing sporadically up to recent times, these studies were primarily concerned with the inheritance of colour in aquarium or pond fish stimulated, presumably, by the 20th Century fashion of keeping fish, especially of tropical species, as decorative pets. Thus although the early work is largely of historical interest, it does illustrate the scope of Mendelian inheritance and did, in fact, accomplish a few 'firsts' in genetics.

3.2 SIMPLE GENE EFFECTS

In fish as in all animals, a uniform body or ground colour is composed of a mixture of pigment types, each developed individually within specific cells. Black pigment, melanin, is the most obvious and occurs within melanophore cells. Yellow pigments within xanthophores are quite common and red pigments in erythrophores are also found (see Goodrich *et al.*, 1941, for chemical identification). One further common epithelial cell in fish is the iridocyte, which is not a true pigment cell but carries crystalline forms of the waste product guanine which acts as a reflecting material giving fish a silvery, or in other contexts a bluish, appearance. Combinations of these pigment cells give rise to a range of body colour types within which the most obvious aberrant form is that of the albino – lacking melanin altogether and leading to a reduction in other pigment types such that the fish look white or pale cream in colour.

Albinism has been observed in a wide range of fish species including cultivated and non-cultivated examples. In the natural environment, examples from primitive fish seem not uncommon. Albino lampreys have been reported several times (Braem and King, 1971) with a possible frequency of one in 100 000 normal fish. In sharks, McKenzie (1970) claimed a first record of albinism in the form of a white hammerhead shark, *Sphyrna levini*, whilst

Talent (1973) listed five records of albino elasmobranchs including one in which the albinos were embryos still in their mother's oviduct and nearly full grown! Talent also observed that albinism is less common in elasmobranchs than in bony fish. Certainly, the flatfish species sought by commercial fishermen frequently throw up examples of yellow or white specimens which get sent in to fisheries laboratories but never arrive alive and rarely get studied further. Albinism in farmed species has been analysed genetically, although not to the extent that it has in aquarium species. It is described as a simple recessive characteristic in rainbow trout (Tave, 1988) and in channel catfish (Bondari, 1981). Gold forms of familiar freshwater fish – tench, *Tinca tinca*, orfe, *Idus idus*, and rudd, *Scardinius erythrophthalmus* – are also common in nature.

Detailed study of the inheritance of albinism in fish largely comes from experiences with freshwater fish bred in captivity. The earliest records include work by Kosswig (1935) with the paradise fish, *Macropodus opercularis*, in which the important distinction is made that body whiteness by itself is not diagnostic of albinism, but that the eyes of the fish must also be without melanin at the back of the retina and hence appear pink (from haemoglobin in blood cells) rather than the normal black. At much the same time Goodrich and Smith (1937) described not only the genetics but also the cellular basis of albinism in the paradise fish. Several other genetic studies with albino fish are referred to by Yamamoto (1969a) in a more recent analysis of albinism in the medaka, *Oryzias latipes*.

Yamamoto obtained his albinos from a population of fish provided by a pet fish dealer. They arose from selected matings of pale coloured but normal-eyed fish and exhibited the familiar features of absence of pigment on the body and in the eyes. Two matings between albino fish generated a total of 800 embryos, all detectable as albinos – this illustrates a useful characteristic of the albino trait, that it can be observed in an individual at late embryonic stages, particularly by the absence of eye pigment which is often prominent in fish embryos. The albino trait clearly 'bred true'.

When Yamamoto cross-mated normal, wild-type medaka to albino, whichever parent was the albino, the offspring were all normal in appearance. When these F_1 individuals were mated together their offspring comprised normal and albino embryos in the ratio 3:1 (actually 423:141, a startlingly good result from the statistical point of view).

The results of these cross-matings are illustrated in Fig. 3.1 together with a summary of the genetic conclusions which can be made. The albino parent in the P_0 generation has the genotype aa (homozygous) comprising an albino allele received from each of its parents. The wild-type genotype is also homozygous, designated AA, these alleles being dominant to the recessive a (capital letters signify dominant alleles although in some of the earlier work this convention was not followed). In the F_1 generation, the offspring are all Aa heterozygotes and normal in appearance but produce eggs and spermatozoa which carry either A or a in roughly equal proportions. Therefore in the F_2, random union of eggs and spermatozoa leads to

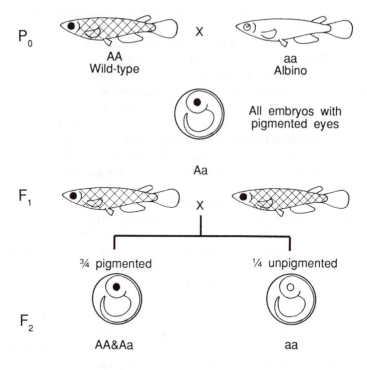

P_0

AA
Wild-type

X

aa
Albino

All embryos with
pigmented eyes

Aa

F_1

X

¾ pigmented

¼ unpigmented

F_2

AA&Aa

aa

Fig. 3.1 The inheritance of albinism in the medaka, *Oryzias latipes*, showing genotypes and associated phenotypes.

the creation of AA, Aa and aa offspring in the ratio of 1:2:1 which, because of the normal appearance of Aa, leads to a ratio of 3 normal to 1 albino embryos.

3.3 COMPLEX EFFECTS OF ALLELES

The exposition of the Mendelian inheritance of the albino gene was a simplification. Yamamoto describes two complicating issues, the first an effect of the albino allele on viability, the second an effect on the interaction of the albino allele with other colour-determining genes.

The viability complication is reasonably straightforward – from the 800 albino embryos only 29 reached adulthood and among the F_2 offspring of other cross-matings the ratio of normal to albino was not 3:1 as in embryos but 20 or more to 1. The albino homozygote aa was much less fit than the AA or Aa genotypes, both of which exhibit the normal wild-type phenotype. This finding was exactly in line with earlier studies of albinism in the Guppy, *Poecilia reticulata*, by Haskins and Haskins (1948), who described the condition as a semilethal mutation. Likewise even earlier work on the swordtail, *Xiphophorus helleri*, by Gordon (1942) reached this conclusion as, more recently, did a study on albino threespine sticklebacks, *Gasterosteus aculeatus*, by Bakker *et al.* (1988). Sundry other studies of albinos in a variety of organisms indicate the effect of the allele on fitness. This secondary consequence of an allele is called **pleiotropy** and is exceedingly common. It is, perhaps, the normal manner of expression of an alternate allele to the wild type. It has important consequences for breeding procedures. Low viability would obviously be disadvantageous for a species subject to domestication, but in more subtle ways any deliberate change to the genetics of a managed species will generate other, correlated responses and these might be undesirable.

The second complication in Yamamoto's albino studies involved genes which control other aspects of pigmentation and these are described in greater detail later in this chapter so, for an example, it is simpler to turn to a study by Haskins and Haskins (1948) on the Guppy. Here too, the albino fish turned up by accident and the condition was demonstrated to be due to a recessive allele which, in the homozygote, prevented the formation of melanin. Our interest here is the interaction of the albino allele with two other melanin genes.

These other genes in the Guppy were first described by Goodrich and his collaborators (1944) and named Gold and Blond. The wild-type background colour of the Guppy is the usual greenish grey characteristic of many fish species, and the two genes represented by recessive alleles in the homozygous condition, gg or bb, both produced a xanthic body colour but by different mechanism. The Gold recessive phenotype showed slightly reduced numbers of melanophores but with a pattern different from that in the wild type, being restricted to the edges of the scales thus giving a more pronounced reticulated pigment pattern against an overall gold coloration. The Blond recessive phenotype on the other hand comprised greatly reduced numbers of melanophores leading to a soft yellow body colour. Both phenotypes included normal eye pigmentation and thus contrasted with the albino phenotype. When Haskins crossed homozygous albino fish with homozygous Blond or homozygous Gold individuals all the offspring were normal. Since these alleles are recessive it follows that the normal F_1 offspring are heterozygotes for both the albino allele and either the Gold or the Blond. In other words the albino allele represents a different gene from the Blond or Gold alleles. Gold and Blond similarly had been shown earlier by Goodrich *et al.* (1944) to be different genes, i.e. non allelic, so we have three genes, all different, generating a reduction in or absence of melanin.

Finally, in the F_2 offspring the different genes segregated independently and some offspring were homozygous for the albino gene and either Blond or Gold recessive genes – these were not intermediate in colour between the albino and Gold or Blond type but indistinguishable from albino. The albino phenotype overrode that of either of the other double recessives – this is termed **epistasis** and is a very common phenomenon in genetics.

3.4 INTERACTION OF GENES

Many of the earlier genetic studies of Mendelian inheritance in fish centred on the domestication and breeding of those species of aquarium fish in which colour variation appeared as a consequence of selective breeding by fish fanciers. Two of the three species of fish have already been mentioned, all appear to have been independently chosen for study and all are classified within a single family of bony fish, the Cyprinodontidae or tooth carps. These species were the Guppy, *Poecilia reticulata*, the medaka, *Oryzias latipes*, and the platy, *Xiphophorus maculatus*. These three species, which still retain their popularity with aquarists, have each had various generic names but ironically, the same common names during their scientific histories!

Colour genetics of *Oryzias*

Oryzias latipes was the first species of fish to interest geneticists. It is a small egg-laying tooth carp from freshwater habitats in Japan. Apart from being easy to rear in the aquarium, it has a decided advantage in genetics in that the eggs are not distributed after spawning but are retained as a clump in a mucous sheath on the ventral surface of the female. Thus if one male is kept with a batch of females, the parentage of eggs is easy to record.

In its natural habitat, the medaka displays no marked variation and fish of both sexes are a dull greenish-grey colour with some flecks of iridescence. Several colour varieties were developed by Japanese aquarists, however, and it is these that formed the basis of the early genetic studies. It is, of course, axiomatic in classical genetics that the inheritance of a character can only be followed if it occurs in recognizably different forms.

The genetics of these colour varieties in the medaka were first detailed by Aida (1921). In contrast to the greenish colour of the normal, wild-type fish, the varieties were either solid red, white or blue and variegated red or white. Aida showed that two different genes were involved. One gene was responsible for the red or variegated red patterns and the phenotypic appearance of the fish was related to the form and distribution of the melanophores. Thus the red phenotype arises when the melanophore cells lack pigment and the normally obscured red pigment (in erythrophores) shows through. The variegated pattern arises when the melanophores are segregated into clumps instead of being distributed fairly evenly over the body. When a red fish was crossed with a wild type, the offspring were all wild type, but when these offspring were crossed, the result was approximately three greenish offspring to one red. The gene for normal development of melanin was therefore given the symbol B as it was dominant over the recessive form of the gene (allele) for no pigment, symbolized b. The crosses are depicted graphically in Fig. 3.2.

The variegated pattern was inherited in a rather different manner. In crosses between red and variegated red, the offspring were all variegated; the allele for variegated was therefore dominant over the allele b and was given the symbol B^1. In crosses between homozygous wild type (BB) and homozygous

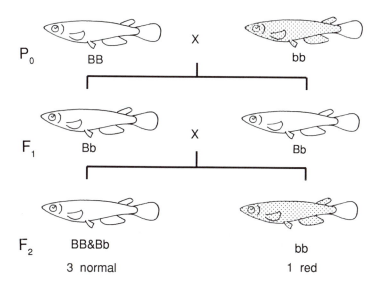

Fig. 3.2 Simple Mendelian inheritance of red body colour in the medaka.

variegated ($B^1 B^1$), the heterozygous offspring were also all variegated but to
a lesser extent than in the homozygous $B^1 B^1$ individuals. Thus B, B^1 and b
represented an allelic series in which b was recessive to the other two which
themselves interacted in a more or less additive fashion and are described as
co-dominant.

The blue and white colours were determined by a different gene from that
of the B series and it displayed quite a different pattern of inheritance. Only
two alleles were present at this second **locus** and they were symbolized R for
the dominant wild-type and r for the recessive allele, the phenotypic consequences of which derived from the absence of red pigment. The two different
colour gene loci interacted at the phenotypic level. Thus with a homozygote
BB:rr, the red pigment was absent and colour was blue due to a dilute showing
of melanin, but in bb:rr, both black and red pigments were suppressed and
the fish were white. This is not the same as albino, where all pigment is
suppressed by alleles at one locus, and was not accompanied by poor fitness.

One of the aquarist strains used by Aida consisted of white females and red
males and bred true. When a red male was crossed with a white female, the
offspring were again either red males or white females. All the fish lacked
melanin, and hence were homozygous for a pair of b alleles, which left the
red male versus white female situation to be explained by the other pair of
alleles. Sex-linked inheritance was well known by the time Aida began his
studies (see Chapter 8) and he based his explanation of the genetics of the red
and white variety of *Oryzias latipes* on the assumption that sex chromosomes
were present: the female had two X chromosomes, the male one X and one Y
chromosome. Thus it was proposed that the allele for white, r, was carried on

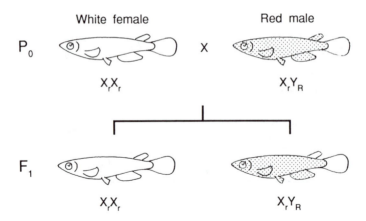

Fig. 3.3 Sex-linked inheritance of red and white body colours in the medaka with postulated sex chromosome genotypes.

the X while the dominant allele R, for red, was on the Y chromosome. The cross is shown in Fig. 3.3. The full interaction of the red and black pigment genes can be seen amongst the second generation offspring (F_2) of a cross between a white female (bb:rr) and a wild-type male ($BB:RR_Y$) where the R allele on the Y chromosome is designated with the subscript Y (Fig. 3.4). The two genes segregate independently into the gametes, sperm or eggs, respectively, and produce a complex array of F_2 individuals which can only be interpreted by careful experiment.

A similar segregation of the variegated patterns occurs if allele B is replaced by B^1 except that the genotype B^1B^1:rr is variegated white, not blue, since the blueness of BB:rr derives from the even distribution of melanophores, not from a blue pigment itself.

The inheritance of the two colour genes in *Oryzias latipes* therefore illustrates both of Mendel's Laws: the discreteness of the hereditary factors, illustrated earlier by the albino gene, and the independent assortment of different genes, touched on in relation to albino and other genes. It also illustrates the existence, however, of multiple alleles (at the B locus), and the phenotypic interaction of alleles at the two loci and finally, the phenomenon of sex-linked inheritance. Thus of the 24 pairs of chromosomes in *Oryzias latipes* (Nogusa, 1960), the R locus is situated on the sex chromosome, and the B locus is on one of the remaining 23 pairs of autosomes. Autosomal versus sex-linked inheritance is a common theme in most animal genetics.

Colour polymorphism in *Poecilia*

The second cyprinodont fish of classical genetic fame is the Guppy, *Poecilia reticulata*, previously known under the generic name of *Lebistes* which is still occasionally used today. The Guppy derived its common name from the Reverend R. J. L. Guppy, who introduced it to Europe as an aquarium fish in

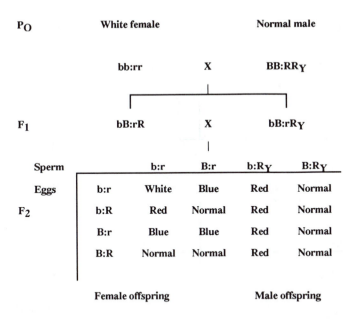

Fig. 3.4 Segregation of alleles at two gene loci in the medaka.

1866. Its natural habitat is in slow-moving streams and pools of South America and the West Indies where populations exhibit a striking polymorphism for colour patterns in male fish. These patterns are highly varied, not only between, but also within populations, and immediately caught the attention of geneticists in the period following the rediscovery of Mendel's Laws. The fish themselves were, and still are, highly desirable animals for experimental genetics. They reproduce readily and bear live young which are robust and easy to rear; they reach sexual maturity within 3 to 4 months under optimum feeding conditions and can, with care, be kept alive for 5 or 6 years.

The first genetic studies on the colour patterns of *Poecilia reticulata* were performed by Johannes Schmidt in 1920 as an investigation of the genetic behaviour of a secondary sexual character. Schmidt was using a breeding stock of the Guppy for experiments on the genetic control of metrical characters, such as the number of vertebrae, and had noted the brilliant coloration of the males as an example of sexual dimorphism. Females were of a uniform grey colour, but the males displayed a specific colour pattern. The coloration in this stock, comprising red and black spots and patches of metallic lustre, was constant from generation to generation but contrasted with a pattern observed by Schmidt amongst stocks at an exhibition of aquarium fish. This observation led Schmidt to the discovery that both contrasting forms were found, amongst others, in wild populations from Trinidad and prompted him to examine the genetics of the different colour forms. The two patterns, which Schmidt labelled A and B, differed in many respects but particularly in the presence of a black spot on the dorsal

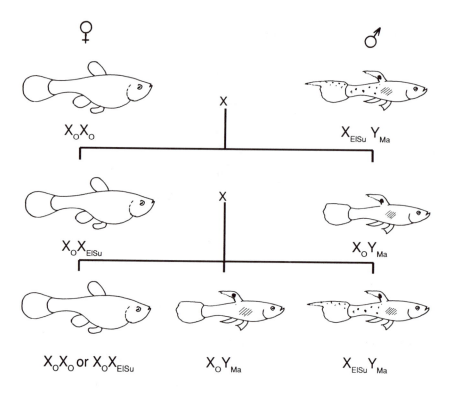

Fig. 3.5 X- and Y-linked inheritance in the Guppy, *Poecilia reticulata.*

fin in B but not in A. A cross between a male of type B and a female from
stock A produced males of type B only, and this result was repeated in several
subsequent generations using B males and their sisters. Similarly, when type
A males were used, male offspring always showed the type A pattern. This
father-to-son inheritance led Schmidt to propose that sex was determined
chromosomally in *Poecilia*, that the male was XY, and that the genetic control
of male colour pattern was located in the Y chromosome.

This hypothesis of Y-linked inheritance was confirmed and amplified by
Winge (1922a,b) with the description of additional male coloration patterns, all apparently linked to the Y chromosome. Winge named these
patterns *Iridescens, Pauper* and *Ferrugineus* and described the patterns as
used by Schmidt as *Oculatus* and *Maculatus*; the latter was the pattern
observed at the aquarists' exhibition typified by a marked black spot on
the dorsal fin. Winge also examined the morphology of the chromosomes
in *Poecilia reticulata* and described the diploid number as 46, later shown
to be 48, proposing that there were 44 autosomes and 2 sex chromosomes,
XY in the male and XX in the female. No cytological difference was evident
between the proposed X and Y but this problem will be dealt with more
fully in Chapter 7.

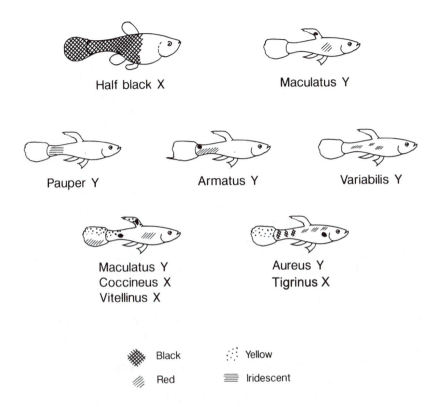

Fig. 3.6 Some of the commoner sex-linked colour patterns of the Guppy and their chromosomal locations.

These were amongst the first examples of Mendelian characters in fish and they also included the first demonstration, along with Aida's work already described, of Y-linked genes. In some respect, however, these male colour genes in *P. reticulata* are exceptional. Males of this species always have some colour spots, and the natural polymorphism is both marked and varied. The individual alleles cannot be distinguished as normal wild-type, versus abnormal or mutant, neither can a **null allele**, i.e. one with no expression, be identified on the Y chromosome since males are always coloured.

In addition to the Y-linked patterns, Winge also observed patterns which were X-linked and showed the conventional sex-linked criss-cross type of inheritance in which genes on the X chromosome are transmitted to offspring of both sexes from the mother but to female offspring only from the father. These were the genes named Sulphureus (Su) and Elongatus (El). In Fig. 3.5 their inheritance is shown to pass from the father to his daughters where they are not actually expressed because they are sex-limited (see p. 36). The F_1 daughters, however, pass the X-linked gene to their daughters or sons and in the latter they are expressed along with whatever genes the Y chromosome contains.

The important points about these examples are first, that the Y chromo-some always has some colour pattern gene – the formula Y_0 has never been observed – and secondly, that although females carry colour pattern genes they do not show the appropriate phenotype – there are exceptions to this which will be mentioned later. These genes are therefore sex-linked and at the same time sex-limited in that they are expressed in males only – they are, in fact, expressed if females are 'masculinized' with male sex hormones. Finally, in the male the genes are always expressed, whether on X or Y or on both, in other words they are dominant genes or alleles.

In his review of the genetics of *Poecilia* Winge (1927) described 18 genes affecting male colour patterns and presented evidence on their chromosomal location. Of the 18 genetically determined colour patterns, 7 were known from the earlier work, 9 were new patterns and 2 were derived by subdivision of 2 of the earlier phenotypes. Some of these 18 patterns are shown in Fig. 3.6. They all show the expected dominance and sex limitation, 9 are linked to the Y chromosome, 8 to the X chromosome and one, *Zebrinus*, is autosomal. Two later additions to this list which do not follow the normal pattern are *Flavus* (X) and *Half Black* (X). Both these patterns, described by Winge and Ditlevsen (1946) and Nybelin (1947), respectively, can be discerned in females and, indeed, in young fish prior to sexual maturation, so they are particularly useful as genetic markers. The patterns, however, are much more vivid in male fish.

Thus 20 different hereditary colour patterns have been described involving at least 20 genes and probably more. Blacher (1927, 1928) had suggested that one of Winge's patterns, *Maculatus*, is controlled by two genes and that another, *Iridescens*, is subdivisible into four components. The genetic basis for these assertions is difficult to evaluate but Winge and Ditlevson (1946) give convincing evidence of the duality of the normal *Maculatus* phenotype. That there must be many more new patterns was suggested by Winge in 1927 and an examination of the wide variety, and splendour, of the Guppy in fish exhibitions today would confirm this view.

These male colour pattern genes, however, are of a very special type in that they are dominant, largely sex-limited and almost all sex-linked. The number of genotypes is certainly very large; the number of gene loci involved is unknown, however, because allelism tests are not possible. Such polymor-phisms are discussed more fully in Chapter 6.

Colour in platyfish

The third cyprinodont fish to figure large in early genetic studies in fish was the platy, *Xiphophorus maculatus*, formerly known under the generic name *Platypoecilus*. This species, like the Guppy, is natural to eastern parts of South America and, like the Guppy, shows a very great deal of natural variation for colour. It also is a live bearer and is easy to rear in captivity. Consequently, it is very popular with aquarists, particularly those with an interest in selective breeding, and has also been studied very extensively by geneticists. For details

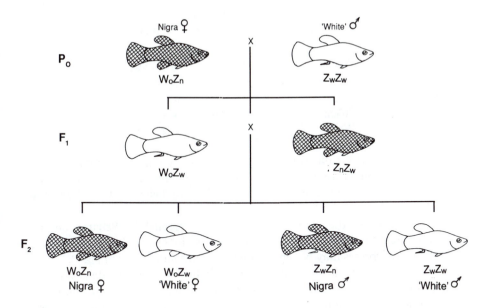

Fig. 3.7 Sex-linked inheritance in the platy – a WZ/ZZ system.

of the care and breeding of these and many other aquarium fish see Axelrod and Vorderwinkler (1957).

Several species of *Xiphophorus* show natural variation (Borowski, 1984) but *X. maculatus* is by far the most variable. The genetic control of its various colour patterns differs from that of *Poecilia* in that most of the genes are not sex-limited and although many of them are sex-linked, they do not form an integral part of the sexual dimorphism in *X. maculatus*. Thus although the Y chromosome in *Poecilia* always carries colour genes, the equivalent chromosome in the platy need not. Colour is not a secondary sexual character in *X. maculatus*. The genetic control of colour patterns is thus of a more conventional type in the platy and allelic relationships can be assessed in a way that is not possible in the Guppy.

The earliest work on *X. maculatus* was performed by Bellamy (1922, 1924), who described the inheritance of a black strain of the fish known to aquarists as 'Nigra'. This character was inherited as a dominant sex-linked allele but, in contrast to the situation in *Oryzias* and *Poecilia*, it appeared that the female was the possessor of the different or contrasting pair of sex chromosomes. Thus a cross between a 'Nigra' female and a lightly pigmented male – described as 'White' by Bellamy – produced 'Nigra' sons and 'White' daughters which, in an F2, produced approximately equal frequencies of 'Nigra' and 'White' individuals of each sex.

Bellamy symbolized the allele for 'Nigra' as N and that for 'White' as W; a departure from the normal convention of symbolizing presumed alleles by a common letter in upper or lower case, depending on dominance/recessive

Table 3.1 Characters and symbols of the genes controlling macromelanophore variation in platies

Bellamy (1924)		Gordon (1927)		Current usage	
Nigra	N	Black	N	Nigra	N
Rubra	R	Red	R	Red	R
Pulchra	P	Spotted	Sp	Spotted	Sp
White	W	Stippled	St	Non-spotted	sp

relationships, with suffixes to identify additional alleles in a multiple series. ('White' probably represents the normal pigmentation in which the melanophores are not concentrated since Bellamy believed that W was the wild-type allele.)

The results of the above crosses were interpreted by Bellamy on the assumption that the differential sex chromosome in the female had a null allele 0. Thus the crosses may be represented diagrammatically as in Fig. 3.7.

This type of sex determination, with a heterogametic female is conventionally described as WZ (female) ZZ (male) system. It is the normal mode of sex determination in birds and some moths and is not uncommon in fish (Chapter 8).

The WZ constitution of the female in X. *maculatus* was confirmed by Gordon (1927) who described in greater detail in this and many subsequent papers the inheritance of a wide variety of naturally occurring pigment patterns. Gordon showed that variation involved three types of pigment cell; micro-melanophores, and macro-melanophores which produce black spotting of various degrees, and erythrophores which produce a red coloration. This red coloration is partly sex-limited, showing more vividly in the male than in the female. The melanophore patterns are expressed equally strongly in males and females. A fourth type of pigment cell, the xanthophore, produces a yellow background coloration and is always present in wild fish.

Initially both Bellamy and Gordon worked with strains of platies obtained from aquarists. These strains had been selected with regard to particular colour characteristics and had been given names describing the colours. Unfortunately, the names adopted by Bellamy were not the same as those used by Gordon and the terminology has become confused, all the more so since possible allelic relationships were not taken into account in allocating symbols to the characters. The strains used by Bellamy and Gordon are listed in Table 3.1.

Gold, an additional character described by Gordon in 1927, lacked both types of melanophores and also erythrophores, thus giving expression to the remaining yellow xanthophore pigments. The pigment cell classification of the five strains as described by Gordon is summarized in Table 3.2.

Gordon (1927) demonstrated that the presence or absence of micro-melanophores was inherited independently of the macro-melanophore and erythrophore patterns which themselves were linked to the W chromosome.

Table 3.2 Chromatophores present (+) in various strains of *Xiphophorus maculatus*

Chromatophore	Strain				
	Red	Nigra	Spotted (Pulchra)	Stippled (White)	Gold
Micro-melanophores	+	+	+	+	-
Macro-melanophores	+	+*	+	-	-
Erythrophores	+	-	-	-	-
Xanthophores	+	+	+	+	+

(*) Extremely abundant

The presence or absence of micro-melanophores was therefore determined by a gene on an autosome. In a cross between a Gold female and a Red male, all the offspring were Red, containing all four types of chromatophore, demonstrating the dominance of genes for the presence of chromatophores over those for their absence. In the F_2 offspring of the cross between these Red males and females, however, the patterns segregated into Red with micro-melanophores

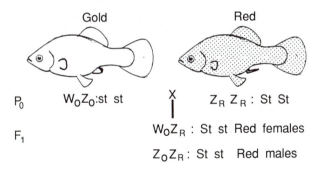

Gold   Red

P_0 $W_O Z_O$: st st   X $Z_R Z_R$: St St

F_1 $W_O Z_R$: St st   Red females

$Z_O Z_R$: St st   Red males

		sperm			
F_2		Z_O St	Z_O st	Z_R St	Z_R st
	W_O : st	St	Gold	RSt	R
Eggs	W_O : st	St	St	Rst	Rst
	Z_R : st	RSt	R	RSt	R
	Z_R : st	RSt	RSt	RSt	RSt

Fig. 3.8 Segregation of micro- and macro-melanophore genes in the platy as shown by F_2 phenotypes. Phenotypes: RSt, Red with micro-melanophores; R, Red; St, Stippled; Gold, Gold.

Table 3.3 Macro-melanophore patterns in *Xiphophorus maculatus*

Name	Symbol	Description	Reference
Non-spotted	sp	Absence of macro-melanophores	Gordon 1927
Spotted	Sp	Irregular blotches on flanks	Gordon 1927
Striped	Sr	Rows of blotches on flanks	Gordon and Gordon 1953
Spotted dorsal	Sd	Irregular blotches on dorsal fin	Gordon and Gordon 1953
Spotted belly	Sb	Blotches restricted to ventral half of body	Gordon and Gordon 1953
Nigra	N	Solid black over most of flanks	Bellamy 1922
Thin nigra	Nt	Extreme striped	Bellamy & Queal 1951
Extended nigra	Ne	Black all over except mid dorsal line	Bellamy & Queal 1951
Fuliginosis	Fu	Random distribution of melanophores giving sooty appearance	Kosswig 1938
Pepper and salt	Sp^1	Very small dots all over	Gordon 1951
Red dorsal	Dr	Red coloration at base of dorsal fin	Kosswig 1938
Black	B	Black all over (?=Fu)	Bellamy & Queal 1951
Micro-pulchra	K	Small dots all over (?=Sp^1)	Bellamy & Queal 1951

(RSt), Red without micro-melanophores (Rst), Stippled (St) and Gold (st). The numbers of offspring of each type approximated to the 9:3:3:1 ratio expected in a dihybrid cross (i.e. 3:1 × 3:1). A schematic representation of the cross is shown in Fig. 3.8 in which the genotype of the Gold female is written W_0Z_0:stst, the Red male Z_RZ_R:StSt. Thus the subscript 0 on the W or Z indicates absence of macro-melanophores and erythrophores, R indicates presence of these two pigment cell types. St represents presence of micro-melanophores, st their absence.

Many different macro-melanophore arrangements have been described in addition to the condition where macro-melanophores are absent. Some of these patterns, together with symbols, discriptions and original references are listed in Table 3.3. Several investigators have expressed the belief that these variations represent a multiple allelic system at a single locus. Thus Bellamy (1928) noted that the patterns R, N, Sp and sp have never been observed in combination in females or more than two at a time in males, even though many attempts had been made to achieve this. In other words, the locus is restricted to the Z chromosome and only one copy is found in females (WZ) and two in males (ZZ). There seems to be no reason why the manifestation of the different macro-melanophore arrangements should not be due to a multiple allelic series, but the inclusion in this of the erythrophore gene R seems unlikely, two quite different cell types being involved. Evidence for the existence of different genes for macro-melanophores and erythrophores was presented by Fraser and Gordon (1928, 1929), who observed the separation of the macro-melanophore and erythrophore patterns in a single fish among the offspring from a cross between a Nigra–Red male and a Gold female. The exceptional fish was a Red, Non-spotted female and as the normal Red phenotype also includes the presence of macro-melanophores, this implies a distinction between R for red and Sp for spotting.

To explain the origin of this anomalous fish, Gordon (1937) proposed that separate genes Sp and R were located on the Z chromosome and had been separated by crossing over, that process whereby homologous chromosomes exchange sections during the period they are closely paired in the first cycle of meiosis. Since the exceptional fish was a female, its Z chromosome was paternal in origin and this presented a problem. The male in question was Nigra–Red, and assuming separate genes for macro-melanophores and erythrophores, could be depicted $Z_{N\,r}\,Z_{Sp\,R}$. Recombining the alleles at the two loci, however, would give either $Z_{\,NR}$ or $Z_{\,Sp\,r}$ which, in combination with the W from the Gold P_0 female, would give Nigra–Red or Spotted females, and the exceptional fish was a Red, Non-spotted female. Gordon therefore proposed that three separate genes were involved, Nigra (N), Red (R) and Spotted (Sp) with corresponding recessives n, r and sp. The P_0 male genotype thus becomes $Z_{N\,rsp}/Z_{n\,R\,Sp}$ and an exchange between R Sp and r sp could give $Z_{n\,R\,sp}$ which with the W from the Gold female would produce a Red, Non-spotted female.

Only one other recombination of this type has been recorded (McIntyre, 1961), and whether the macro-melanophore/erythrophore patterns are due to one locus or to a complex of genes is still to be decided (Kallman and Atz, 1966).

More extensive data on crossing over (Chapter 7) in *X. maculatus* are given by Bellamy and Queal (1951), not on the basis of recombination within the macro-melanophore system, but between it and the chromosome region concerned with sex determination. On the assumption that the W and Z chromosomes are largely homologous and differed only with respect to a 'differential sex-segment', it is possible to conceive of recombination between the colour genes and sex. Bellamy and Queal summarized data on two sorts of crosses:

(a) $W+Z_c$ female $\times$ $Z+Z+$ male
(b) $W+Z_c$ female $\times$ Z_cZ+ male

where c stands for any of the macro-melanophore colour patterns and + for no colour. In (a) the offspring should comprise coloured males and uncoloured females but amongst 2679 females, 13 (0.49%) were coloured and amongst 2734 males, 14 (0.51%) were not coloured. In the crosses of type (b), the situation is more complex and a Wc could be recognized in those female offspring which received a Z_c from the male, and a Z_+ could be recognized in those male offspring which received a Z_+ from the male parent. Of 2692 female offspring 9 (0.33%) were recognized as exceptionals and 6 (0.23%) out of 2596 males – roughly equivalent to half the frequency of exceptionals in crosses of type (a).

Bellamy and Queal subjected some of the exceptionals to further breeding tests and listed four outcomes:

1. exceptional males which proved to be genotypic females (WZ) not cross-overs; such sex reversals are not uncommon in fish;
2. fish in which the new combinations proved stable and showed normal Mendelian inheritance, i.e. genuine cross-overs;

3. cross-overs which produced a high proportion of WZ males in the next generation;
4. sterile fish – these were assumed to represent non-disjunction, the failure of a homologous pair of chromosomes to separate during meiosis in P_0 fish leading to individuals with an extra chromosome or with one missing in the F_1.

Roughly half of the exceptional genotypes in the F_1 offspring therefore appear to be recombinations between that part of the chromosome concerned with sex determination and that part containing the macro-melanophore genes. Thus the recombination frequency is about 0.25%. This is a rather low figure and implies that the colour genes and sex determinants are close together on the sex chromosomes. This close association is unlikely to be a random event. With 24 pairs of chromosomes, the chance that any two genes at random are on the same chromosome is 1 in 24, and the chance of random association of a group of genes into one small region of a chromosome is a great deal less likely than this. Thus the macro-melanophore–sex complex has been referred to as a super-gene, or switch gene with a functional association in relation to sex and reproduction. (See also Chapter 8.)

3.5 SUMMARY OF COLOUR GENETICS

Thus we see that colour in fish can be determined by the action of genes with alternate alleles. Control may be easily demonstrated if a simple dominance/recessive relationship holds but this is not often the case. Where complex pleiotropisms or epistatic interactions occur, further complicated by multiple genes and multiple alleles plus the complications of sex chromosome inheritance, the situation may be enormously variable. Under these circumstances the resolution of the genetic control could only come about by very careful construction of cross-matings and diligent analysis. Such studies are rarely performed in modern genetics and much of the huge range of variation seen in the colour of pet fish and also in some farm fish remains of unknown genetic derivation.

Golden rainbow trout, for example, have been bred at fish hatcheries for many decades but incontrovertible evidence of the genetic control of colour in this ubiquitous fish is hard to find. In a popular review, Tave (1988) describes albinism in rainbow trout as due to an autosomal recessive allele. Golden and Palomino phenotypes are described as due to two alleles with additive effects – genotypes GG and G'G' are Normal and Gold, respectively, whilst GG' shows an intermediate phenotype, the Palomino. These latter fish are said to be popular with sport fishermen for the variety they introduce. Strict proof of these modes of inheritance is difficult to find and further claims on the inheritance of blue coloration in rainbow trout by Tave (1988) is even more conjectural. In Chapter 14 we shall see that an enormous diversity of colour is characteristic of a range of modern ornamental fish species, and the genetic control of it must largely rely on extrapolations from the old literature.

Returning to the aquarium species, such difficulty with something as visible as body colour or pattern should give pause to the thought that we can make simple assertions about the genetics of other aspects of the form or function of fish.

Chapter four

Mendelian inheritance: II. Characteristics other than colour

4.1 INTRODUCTION

The overwhelming majority of studies in fish on the inheritance of simple Mendelian traits has been with body colour or colour pattern. Most of this work has involved the use of aquarium species as opposed to fish of commercial importance for food. The reasons for these obvious developments are clearly that colour and pattern are easy to see and that breeding studies are facilitated by the short generation time in aquarium fish. The popular aquarium livebearers can be sexually mature within 6 months of birth whereas carp and trout may require 3 years to reach sexual maturity and salmon and halibut may take several years more than this.

Characteristics other than colour have received some study – body shape, scale and fin characteristics are important, both for the pet fish breeder and for the commercial fish farmer, and sundry other traits have received sporadic attention. These will be covered briefly in this chapter together with the inheritance of biochemical characteristics which can be revealed by the techniques of electrophoresis. These characteristics have been studied in a wide range of fish species throughout the world and published works on this topic must out-number all other genetic studies in fish. They are, however, overwhelmingly devoted to the identification or analysis of populations. Only very limited efforts have been made to assess the inheritance, in purely Mendelian terms, of these relatively simple characteristics.

4.2 SCALE PATTERNS OF CARP

The classical example of inherited variation in scale patterns is that of the various forms of common carp, *Cyprinus carpio*, used in farming and for recreational fishing. Four basic types of carp are recognized in this context:

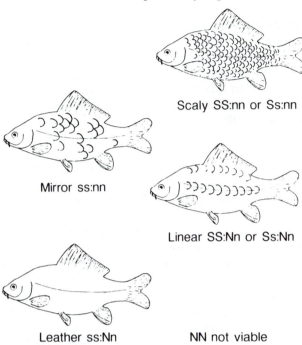

Scaly SS:nn or Ss:nn

Mirror ss:nn

Linear SS:Nn or Ss:Nn

Leather ss:Nn   NN not viable

Fig. 4.1 Scale pattern phenotypes in the common carp *Cyprinus carpio*, (TL about 40 cm).

1. scaly carp, frequently described as 'wild' carp;
2. mirror carp, with a few large scales scattered about the body;
3. linear carp, in which the normal overlapping tiled effect of the scales is transformed into linear arrays running the length of the body;
4. the leather carp, with very few scales at all.

These features are illustrated in Fig. 4.1.

These variations have been given 'variety' status by some farmers and fishermen. They seem to have been developed in European carp culture long before the advent of genetics but Twentieth Century genetic study has shown that they all derive from the effects of only two genes. Varietal and higher levels of taxonomic separation must involve hundreds, possibly thousands of differing genes so these scale patterns are not of taxonomic significance.

The simpler of the two genetic systems from which these scale phenotypes arise generates the mirror form. There are two 'scaly' alleles at a locus designated S and s and the mirror carp is homozygous for s; i.e., the genotype is ss. The allele s is fully recessive, so that genotypes Ss and SS are indistinguishable and fully scaled. The second gene also has two alleles but the one producing the linear form of carp is dominant over the recessive normal allele. The heterozygote Nn is a linear carp, the recessive homozygote nn is normally scaled and the dominant homozygote NN is not viable, dying during the embryonic phase of development. The fourth of the four forms, the leather carp, is the combination of the two phenotypes, mirror and linear, and such fish have the single genotype ss:Nn.

Because of the heterozygous nature of the linear genotype, linear and leather carp do not breed true. Linear carp crossed together (Nn × Nn) produce offspring of which one-third are homozygous for the normal allele (nn) and are therefore scaled or wild-type in appearance. The other two-thirds are heterozygotes (Nn), and hence show the linear scale patterns, and NN individuals do not survive. The same segregation at the N gene occurs in the leather × leather cross, but interaction with the s gene, where only one allele is represented, generates one-third mirror carp and two-thirds leather. Crossing of linear × leather produces either linear plus scaled fish or linear scaled and mirror carp, depending on whether or not the linear fish were heterozygotes.

All of these forms of carp are found in farm and fishery use world-wide and little note is taken of the fact that there is a built-in mortality of 25% in any cross between linear and/or leather carp. Most of the very large fish which are reported by anglers from sport fisheries in the United Kingdom appear to be either mirror or leather in appearance. There is no scientific evidence that they grow any quicker or live longer in fisheries than the scaled form. It remains possible, however, that the frequent handling which arises when such fish are caught and returned to the water causes more damage to the scaled than to the mirror or leather type and that mortality of the former is higher under sport fishery conditions.

Most of the photographs of large specimens also appear to show minimal scaling even when the fish are described as mirror carp. Within each of the non-scaled forms there is, however, considerable variation of expression of the characteristic. This is due in part to so-called modifier genes which enhance or diminish the effect of the 'main' allele. Modifier genes and quantitative inheritance will be dealt with in Chapter 5.

Other apparently scaleless phenotypes are commonly observed in the colourful variety of goldfish called shubunkins. This is not genuine lack of scales, however, but a lack of the silvering material guanine, leaving the scales transparent. In the shubunkin this characteristic is reported to be due to a dominant allele in the homogygous state (Zhong-ge, 1984). The homozygous recessive allele is the wild-type metallic form and the heterozygote is claimed to be intermediate or nacreous.

Other authors claim more complex hereditary patterns in the shubunkin, with two pairs of genes showing epistatic interactions (Matsumoto *et al.*, 1960). These details will be returned to in Chapter 14.

Similar characteristics of scale transparency are found in common carp, but according to some authors the inheritance is simple with a dominant wild-type allele and a fully recessive mutant allele which produces the transparent phenotype only in the homozygote (Chen, 1928; Probst, 1949). Others, for example Vanderplank (1972), claim that some of the heterozygotes show a patchy absence of guanine which would resemble the nacreous appearance of goldfish heterozygous for the allele.

4.3 SCALE PATTERNS IN STICKLEBACKS

The scale variations of common carp were probably developed deliberately by carp breeders from the chance occurrence of mutant forms during domestication. A somewhat similar expression of an extreme scale pattern variation exists naturally in populations of the threespine stickleback, *Gasterosteus aculeatus*. This tiny but elegant little fish is found in fresh water, brackish and marine waters over much of the Northern Hemisphere. One of its principal claims to fame arises from its complex mating displays and the parental care of eggs and fry exercised by the male. Much fish behavioural study has been based on these characteristics (Wootton, 1976). The other important feature of these fishes is the dramatic scale polymorphisms which they exhibit. Their scale covering is quite unlike that of the 'roof tile' pattern normally seen in fish, but comprises an array of vertical armour-like plates arranged in a single row along the body. The three distinctive forms, as found in northern European localities, are illustrated in Fig. 4.2 and named in accordance with the terminology first used by Bertin (1925) in a study of their geographic distributions in and around the North Sea.

In order of increasing levels of scalation, the least scaled has only four or five lateral plates at the front end of the body, the rest of which is naked. Bertin called this form *gymnura*. The next level of scaling comprises plates over about half of the front end of the body with a small, horizontal keel in the mid-lateral part of the slender caudal peduncle; this form was called *semiarmata* by Bertin and has 7 to 15 plates. The most extreme scaled form is completely clad in bony plates and the keel is well defined resembling those typical of fish of the tuna family. With up to 35 plates, this form was termed *trachura*.

Subsequent investigations of the scale patterns of sticklebacks have used sundry other terms to describe the various forms or morphs. The name *leiura* seems to have replaced *gymnura* and an extreme form with no scales at all has been called *hologymnura*: it has not figured much in subsequent study, possibly because of its very limited distribution, appearing only at the southern limit of the species in the Mediterranean region. Other complications include the use of the feminine of Bertin's terms, i.e. *trachurus*, etc., but the greatest departure is the American use of *complete-partial-* and *low-plated* to describe the three common forms (Hagen and Gilbertson, 1972). A further confusion arises over the use of *trachurus* and *leiurus* to define marine and freshwater populations. All of these issues and other historical aspects of the stickleback polymorphism have been reviewed by Bakker and Sevenster (1988) who recommend use of the original masculine terminology *trachura*, *semiarmata* and *leiura* for the plate forms and anadromous and freshwater to describe the different populations or life styles. These simple and logical recommendations will be followed here.

Bertin's work, reported in 1925, established that the three forms were not different species as suggested by earlier workers, but interbreeding populations in which the degree of scalation was correlated with geographical locality and with

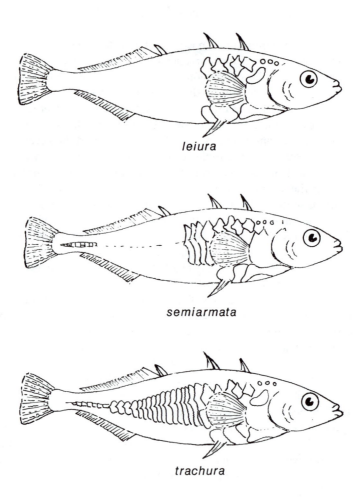

leiura

semiarmata

trachura

Fig. 4.2 Plate morphs in the threespine stickleback, *Gasterosteus aculeatus* (TL about 3 cm).

salinity. Thus the *trachura* form was predominant in northern marine localities whilst *leiura* was typical of southern and freshwater environments. Bertin concluded that the scale characteristics were determined by temperature and salinity in a Lamarckian sense.

This idea did not receive much support and in any case seems inconsistent with the discontinuous nature of the variation and the fact that all of the forms breed in fresh water or brackish water where the fry spend their early formative months. The Lamarckian explanation was eventually set aside by breeding experiments which showed that crosses between *trachura* and *trachura* produced only *trachura* whatever the environmental conditions and, similarly, *leiura* forms bred true. The first work of this nature was by Heuts (1947) on Belgian populations of sticklebacks. This was one of the earliest attempts

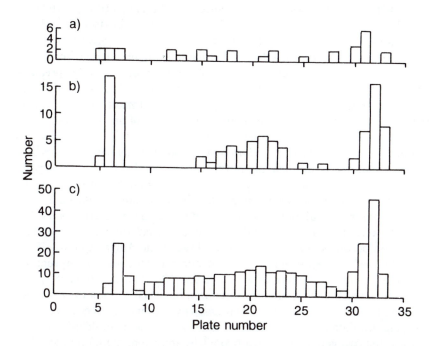

Fig. 4.3 The distribution of plate numbers amongst offspring from crosses involving *semiarmata* forms of the stickleback. Note varying vertical scales, (a) a backcross with *trachura*, both parents from inter-locality crosses; (b) an F_2 where the P_0 comprised *leiura* and *trachura* of different localities; (c) crosses between *semiarmata* from one locality. (After Münzing, 1959).

to define the evolutionary significance of polymorphisms by experiment, and the demonstration of the increased survival of *trachura*, as opposed to *leiura*, on transfer from fresh water to salt water gave credence to the conclusion by Heuts that the plate morphs were genetically determined. The evolutionary significance of many other factors, including temperature, salinity and susceptibility of the stickleback to predators or disease, has been reviewed by Bell (1984), who concluded that potential selective forces could be recognized in many natural situations but that a convincing account of the total process of natural selection for plate morphs could not yet be given. The high level of management applied to many aquatic ecosystems today, particularly of freshwater reservoirs, ought to give abundant opportunity to carry these studies further forward.

To return to the genetic control of the three plate morphs or phenotypes in sticklebacks, Heuts (1947) only differentiated between two, *trachura* and *leiura*, and regarded *semiarmata* as a hybrid between these. Heuts concluded that genetic control was by polygenes (Chapter 5), i.e. that it was multifactorial and not due to Mendelian inheritance of major genes. Subsequent study has shown that both forms of inheritance are implicated. A far more comprehensive set of crosses was performed by Münzing (1959, 1962) on sticklebacks of European origin and by Hagen and Gilbertson (1973) on American threespine sticklebacks. Münzing made crosses between all of the possible forms, and confirmed that *trachura* and *leiura* bred true whereas the *semiarmata* × *semiarmata* cross produced offspring with a wide spread of plate numbers (Fig. 4.3) with peaks corresponding to *trachura* and *leiura*, respectively, but with a wide band of possible *semiarmata* forms. Münzing concluded that this could be explained on the basis of one gene locus with two alleles showing a dominant/recessive relationship. Thus *trachura* was produced by the genotype TT, *leiura* by the homozygous recessive tt whilst the broad band of *semiarmata* phenotypes corresponded to Tt with varying levels of genetic modification. In this context it should be noted that the *trachura* and *leiura* crosses were 'pure', i.e. the males and females came from the same natural population. The *semiarmata* × *semiarmata* crosses used F_1 parents derived from cross-matings between *trachura* and *leiura*. This distinction becomes important when we look at the American data.

Doubt has been expressed over the adequacy of Münzing's interpretation of simple Mendelian inheritance (Wootton, 1976) and Bell (1984) has attempted a reinterpretation of the data on a model embracing the existence of two genes. In this model, initially proposed by Hagen and Gilbertson (1973) to explain the data from their genetic studies with American stickleback plate morphs, the two genes each have two alleles (one dominant, one recessive) and the plate morph types are produced by different ratios of dominant and recessive alleles with no other form of interaction. This is akin to quantitative inheritance which will be described in the next chapter. The consequences of various combinations in this digenic model are illustrated in Fig. 4.4 on the basis that the *trachura* form is produced when the genotype has one recessive allele or none, the *semiarmata* form from a genotype with any two

locus A genotype

		AA	Aa	aa
	BB	*t*	*t*	*a*
locus B genotype	Bb	*t*	*a*	*l*
	bb	*a*	*l*	*l*

Fig. 4.4 A digenic, non interactive model for plate morph patterns in the stickleback (after Bell, 1984). Phenotypes: a, *semiarmata*; l, *leiura*; t, *trachura*.

		trachura		
		AABB	AaBB	AABb
	AABB	*t*	*t*	*t*
trachura	AaBB	*t*	*t/a*	*t/a*
	AABb	*t*	*t/a*	*t/a*
	AAbb	*t*	*t/a*	*t/a*
semiarmata	aaBB	*t/a*	*t/a*	*t/a*
	AaBb	*t/a*	*t/a/l*	*t/a/l*
	aaBb	*t/a*	*t/a/l*	*t/a/l*
leiura	Aabb	*t/a*	*t/a/l*	*t/a/l*
	aabb	*a*	*a/l*	*a/l*

Fig. 4.5 Plate morph phenotypes generated by some of the crosses within the digenic model of Fig. 4.4. Two further series would embrace *trachura* × *semiarmata* and *trachura* × *leiura* crosses.

alleles of the recessive form and the *leiura* form when three or four recessive alleles exist in the genotype. A consequence of this model is that pure-breeding or segregating crosses can be obtained from any combination. For an *armata* type AA bb will breed true but the same phenotype with the double heterozygote Aa Bb will not. This is a very flexible model but difficult to test predictively in the way old-time Mendelianists would approve of, but more significantly, there does not appear to be anything quite like it in other genetic studies of major gene effects.

Bell (1984) points out that of the 18 crosses performed by Münzing (1959) only 6 fit the single locus model without fairly drastic assumptions such as poor penetrance of the dominant allele T, i.e. that it is frequently not expressed in fish carrying it. If the digenic model is applied to Münzing's data the anomalies become far fewer. Figure 4.5 depicts some of the crosses of the three possible genotypes for each of the plate morph phenotypes under the digenic model. There are 9× 9, i.e. 81, different matings possible within this model and it is a simple exercise to deduce that there are 12 different segregation pattern possibilities ranging from monomorphic to a 5:6:5 segregation for *trachura*, *semiarmata* and *leiura* plate morph patterns, respectively, and even including the asymmetric patterns 1:3:4 and its mirror image. Clearly, with an acceptance of some statistical variation, almost any observed segregation pattern can be explained by the digenic, non-interactive model.

Although the plate morphs in North American sticklebacks look indistinguishable from those in European fish, there are differences in distribution. Thus in Europe the monomorphic populations are either *trachura* in marine localities or *leiura* in fresh water with dimorphic and trimorphic populations apparently in areas where mixture of marine and fresh water can occur. In contrast, the North American populations include monomorphic populations of *semiarmata* type as well as of the other two. This immediately throws into doubt the notion of simple, single locus inheritance since segregation would rule out *semiarmata* monomorphic populations. In addition to the monomorphic populations some exist where all three types can be found and sometimes in situations such as land locked lakes where interchange between marine and freshwater populations cannot take place.

Hagen and Gilbertson (1973) demonstrated that all three types bred true when taken from monomorphic populations even when the two parents did not come from the same population. When two different forms from monomorphic populations were crossed, the results depended on the origin of the parents. Thus in one cross of *trachura* male × *leiura* female, and its reciprocal, all offspring were *semiarmata,* just as in the European crosses, whereas *trachura* × *leiura* reciprocal crosses, using *leiura* from a third population, produced only *trachura.*

The analyses of F_2 crosses seemed only to use *semiarmata* of the above F_1 crosses and those segregated in an apparently normal 1:2:1 Mendelian single locus model with no dominance. Of greater interest would have been F_2 segregations from the F_1, *trachura* individuals which came from the *trachura* × *leiura* cross since such individuals would express segregation patterns too.

The other sorts of crosses analysed by Hagen and Gilbertson (1973) came from a single population (Lake Wapata, Washington State) which included all three plate morphs. Four crosses involving only *semiarmata* all segregated for all three plate morphs with no conventional ratio but approximating to equal frequencies of *leiura* and *semiarmata* with *trachura* about 30% less than these. Crosses (26) between *leiura* parents mostly generated only *leiura* offspring but in three cases there was segregation for *leiura* and *semiarmata* forms, with the former predominating. No crosses involving only *trachura* were reported but in two between this form and *leiura* one segregated more or less equally for the parental types whilst the other produced all three forms at about equal frequencies. Finally, 23 crosses between *leiura* and *semiarmata* produced one non-segregating *leiura* family whilst the others segregated for *leiura* and *semiarmata* or for all three morphs with varying ratios. Hagen and Gilbertson (1973) argued that these ratios could not be explained by a single-locus model but could fit a two-locus model as explained earlier. They described a goodness-of-fit test for this model in all their results and concluded that they were compatible with the model.

One might go back to the crosses involving individuals from monomorphic populations for a simple test which might check the model. When *trachura* of one population was crossed with *leiura* from population B and C, one produced only *semiarmata*, the other only *trachura* offspring. There seems to be no combination of parents from the model (Fig. 4.5) which can accommodate these observations. Of the three possible *trachura* genotypes and three *leiura* only one gives all *semiarmata* offspring, none give all *trachura*.

Further work on land-locked North American populations by Avise (1976) may provide the clue to resolve the plate morph inheritance paradoxes. A freshwater population in California (Lake Friant) contained fish with plate numbers around the modes of 7 or 31: these can be equated to *leiura* and *trachura* respectively; the *semiarmata* form was virtually absent. Crosses between the two forms produced offspring of one or other form or of both; only 1.5% of the offspring were of intermediate plate count. When similar crosses were done using Lake Friant fish but with the other parents coming from either of two quite different populations, a similar picture emerged but with the essential difference that the frequencies of intermediates were much higher than for the straight land-locked crosses, at 12.1% and 25.0% respectively. Avise argued that the total genic array in the Lake Friant population could be 'co-adapted' to favour production of only the two distinctive morphs but that destruction of this co-adapted gene pool by 'hybridization' could disturb the control mechanism and allow the intermediates to be expressed phenotypically. This problem of 'co-adapted' gene pools is of considerable importance in connection with fish farming and the risk that escaped fish could damage natural populations. Further discussion on it will continue in Chapter 9.

To summarize on the stickleback plate morphs, the evidence suggests that major genes exhibiting Mendelian inheritance are probably involved, but that many small genetic modifying factors may influence the expression of the characters and that the genetic background may also impose further constraints on body scale patterns. We may therefore return to the initial proposal by Heuts (1947) that plate number is under control by polygenic inheritance and represents a quantitative genetic trait (Chapter 5).

4.4 BODY SHAPE

Most of the scientific studies of the genetic control of body shape do, in fact, deal with skeletal abnormalities, particularly those involving the spine. Specific Mendelian inheritance was described by Aida (1930) for two different abnormalities of the spine in *Oryzias latipes* (medaka). They were non-allelic genes and the recessive alleles were called *wavy* and *fused*, respectively. *Wavy* in the homozygous condition caused a curvature of the spinal column in the vertical plane and in consequence reduced the length of the body relative to its depth. The allele *fused* produced a similar body shortening by reducing the length of the spinal column. Studies of the inheritance of the two

genes suggest they are on different chromosomes (Takeuchi, 1976). Similar
shortened body characteristics related to spinal deformity have been reported
in other aquarium species. In the Guppy, Goodrich *et al.* (1943) describe a
mutant form similar to *wavy* in the medaka and it too was determined by a
recessive allele. Rosenthal and Rosenthal (1950) described two similar conditions in the Guppy of different origin, one of which was inherited as a simple
Mendelian recessive and called lordosis. The lordotic mutant allele was also a
semilethal in the sense that segregation ratios were heavily biased to the wild
type and against the homozygous recessive. In fish farmed for food such
mutations would be extremely disadvantageous. They do occur sporadically,
however, of environmental and genetic origin (Fig. 4.6 (a) and (b)).

The second of the spinal deformities in the Guppy described by Rosenthal
and Rosenthal (1950) was shown not to be inherited. This is probably true
of most such cases in fish and the cause can be attributed to environmental
accidents, particularly around hatching time. One genetic form that might be
significant, however, arises as a consequence of inbreeding (Chapter 6). In
salmon, McKay and Gjerde (1986) describe skeletal deformities of this sort but
interpret the data as indicative of genetically enhanced susceptibility to
nutritional deficiency – vitamin C is well known to be implicated in such
conditions. In farmed rainbow trout (Purdom, unpublished), very significant
genetically controlled abnormalities of body shape could be identified in an
inbred strain and such occurrences in inbred birds and other animals are well
known (Lerner, 1954). In an inbred group of aquarium fishes, *Poeciliopsis
prolifica*, abnormality of the vertebral column did, in fact, turn out to be
inherited as a Mendelian recessive (Schultz, 1963). Its viability was low,
however, since male fish were rarely able to copulate successfully. This
illustrates a general principle that in any breeding programme that entails
some level of inbreeding, rare recessive mutant forms will turn up, often
showing severe fitness defects (Chapter 6).

The area of fish breeding where developments in body shape are obvious
and, at times, dramatic is in the production of fancy goldfish, *Carassius auratus*.
The shortened, almost lordotic bodies of such strains as the fantail goldfish,
the shubunkin and, in an extreme form, the veiltail and lionhead, are very
similar in appearance to skeletal defects of other fish (Fig. 4.6 (c)). These fancy
forms seem to breed true and their characteristics must therefore be controlled
by major genes. The fancy strains all emerged long ago and well before the
conscious use by breeders of Mendelian principles. Furthermore, there is still
very little information on the specific mode of inheritance of such traits. It
seems probable, however, that each of them represents the impact of a major
gene further developed by deliberate selection by breeders to enhance the main
effects of the gene. As with plate morphs in sticklebacks, a combination of
major and polygenic inheritance is the most reasonable explanation of the
array of distinctive body shapes.

Some aspects of body shape in goldfish have been attributed to individual
genes. The best example is the protruding eye characteristic called *telescopic* which
Matsui (1934) attributes to a single gene, but there can be little doubt that its

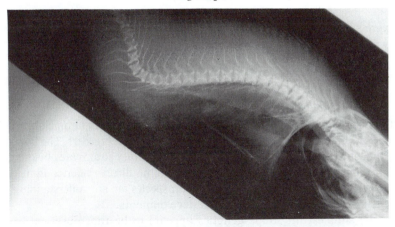

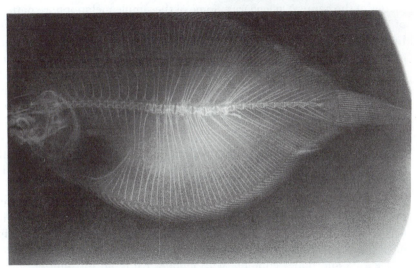

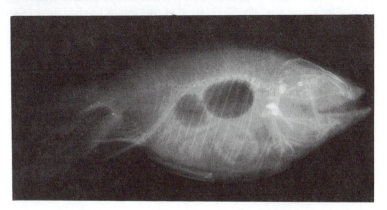

Fig. 4.6 Spinal deformity in the cod (a), sole (b) and goldfish (c) as revealed in X-ray photographs (TL about 60, 18, and 12 cm respectively).

expression in such forms as 'bubble eye' have been enhanced very significantly by selection.

4.5 FIN SHAPE

The ubiquitous goldfish once again represents a principal example of variations, this time of fin shape, developed by generations of fish breeders. The flowing fins of the comet, veiltail, fantail moor and most other fancy strains are obvious genetic developments. As with body shape, we have no direct evidence from goldfish studies and must turn to other aquarium species for an understanding of the genetics behind these often beautiful developments.

The simplest fin shape genetics appears to be that observed in fancy platies of the genus *Xiphophorus*. Some of the simple melanophore patterns of *X. maculatus* were described in Chapter 2 but colourful varieties of varied hues have been developed and given exotic names such as *bleeding heart, gold-tuxedo* and *red-wagtail* (Gordon, 1951). Within these strains, developments in the 1960s led to fancy forms with a very pronounced dorsal fin – called *topsail* or *hifin*. This characteristic was shown to be inherited as a Mendelian dominant (Norton, 1967a), and as a dominant it was easy to introduce to sundry other colour forms. In the closely related swordtail, *X. helleri*, the same *hifin* characteristic is known, again a dominant characteristic but, in addition, a similar fin elongation pattern is expressed in the tail fin producing a lyre-shaped caudal fin, hence the name *lyretail* in this species. This too is inherited as a dominant (Norton, 1967b). The two characters are due to different autosomal dominant alleles and in combination produce the very ornate *hifin lyretail* combination (Fig. 4.7) which, according to Norton (1967b), is easily recognized in very young fish but associated with sterility in the male. Fish of the genus *Xiphophorus* are all livebearers and males copulate with females using an intromittent organ developed from the anal fin. Elongation of this might be the cause of sterility but details are not available. Similar sterility occurs in male *Lebistes reticulatus* of the veiltail breeds – to be described below – but this is behavioural and

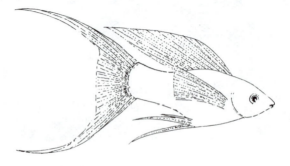

Fig. 4.7 The *hifin lyretail* platy (TL about XX cm) (after Norton, 1967b).

when the 'excess' tail is removed surgically the male can successfully mate. A mutation (Lyra) in *Mollienesia lattipinna* produces elongation of all the fins (Schroder, 1964). This too is dominant, but unlike the situation in *Xiphophorus*, the homozygote for the Lyra allele, LyLy, is inviable.

The veiltail or delta tail Guppy expresses fin shape variation to extreme levels, with the caudal fin in good male specimens exceeding in length and depth the length of the rest of the body by considerable amounts. Schroder (1974) describes a gene *Kalymma* which generates long flowing fins in *Lebistes reticulatus* but its complex pattern of inheritance requires the postulation of a suppressor gene. Thus *Kal* produces flowing fins only if a suppressor allele is absent. In my experience, the outcrosses and subsequent generations of offspring from a veiltail × normal Guppy cross generate a complex mixture of fish with various expressions of fin shape including small veiltails, normal fins, strap-shaped tails, upper swordtails, lower swordtails and lyretails, all of which can be seen in individual strains of these popular fish. A complex mixture of dominant genes is the simplest hypothesis but the mechanism by which they were brought together in the veiltail is an example of a modern triumph of unconscious genetic selection. Foerster (1987) describes the recent history of the development of two sorts of veiltail Guppy. One was typified by elongation of all fins in the male including the gonopodium, which effectively sterilized the males. The other strain (Greissen) had much the same appearance but the fin characteristic was expressed in females as

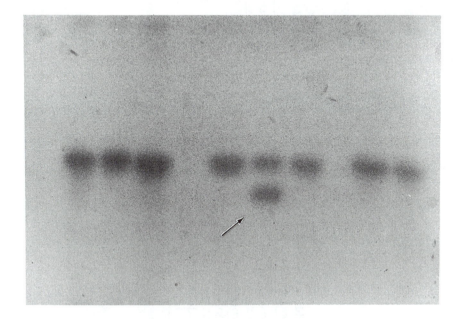

Fig. 4.8 Electrophoretic appearance of the rainbow trout serum enzyme phosphoglucomutase; a simple monomer showing 7 fish homozygous for a fast allele and one heterozygous for fast and slow alleles (arrowed).

Table 4.1 Segregation of offspring in conventional crosses made at the Pgm locus: numbers in parentheses are the expected segregations from Mendelian ratios

Parental genotypes (female first)	No. of matings	Offspring genotypes			χ^2	Probability
		1.00/1.00	0.85/0.85	1.00/0.85		
1.00/1.00 × 1.00/0.85	4	141 (146.5)		152 (146.5)	0.41	0.70>0.50
0.85/0.85 × 1.00/0.85	2		34 (42.5)	51 (42.5)	3.40	0.10>0.05
1.00/0.85 × 1.00/1.00	9	537 (559.5)		582 (559.5)	1.81	0.20>0.10
1.00/0.85 × 1.00/0.85	2	71 (69.25)	51 (69.25)	155 (138.5)	6.82	0.05>0.02

Fig. 4.9 Electrophoretic appearance of the rainbow trout serum enzyme superoxide dismutase; a dimer, showing 3 alleles, giving homozygotes for the fast allele (2 fish), for the medium allele (7 fish), heterozygotes for fast and medium (5 fish, one of which is arrowed) and heterozygotes for the medium and a slow allele (2 fish, one of which is double-arrowed).

well as in males. The fin elongation was attributed to a dominant gene with more extreme expression in the homozygote. The significant aspect of these German developments in the veiltail Guppy is simple expression – all fins elongated. In several of the photographs in Foerster's paper, the veiltail Guppy like those commonly seen in aquarist shops do not have elongated fins other than the caudal.

4.6 ELECTROPHORETIC VARIATION

Electrophoresis entails the study of the mobility of soluble proteins across an electric potential. The basic methodology employs a matrix, such as a sheet of agar gel, which may be buffered to a specific pH and to which is applied an electric potential. The material to be analysed is placed in a slit in the gel between the two electrodes and specific proteins are subsequently stained to reveal the position to which they have migrated under the influence of the electric potential.

For a limited number of proteins, genetic analysis has shown that a very simple Mendelian type of inheritance arises in which an individual band, or a specific cluster of bands, is determined by an allele and that the combined effect of two different alleles is simply their sum. Thus there is no dominant/recessive relationship in electrophoresis and the alleles are described as co-dominant.

Table 4.2 Segregation of offspring in conventional crosses made at the Sod locus: numbers in parentheses are the expected segregations from Mendelian ratios

Parental genotypes (female first)	No. of matings	Offspring genotypes			X^2	Probability
		1.52/1.52	1.00/1.00	1.52/1.00		
1.00/1.00 × 1.52/1.00	2		126 (134.5)	143 (134.5)	1.07	0.30>0.20
1.52/1.00 × 1.00/1.00	9		341 (342)	343 (342)	0.01	0.95>0.90
1.52/1.00 × 1.52/1.00	5	61 (59)	60 (59)	115 (118)	0.16	0.95>0.90

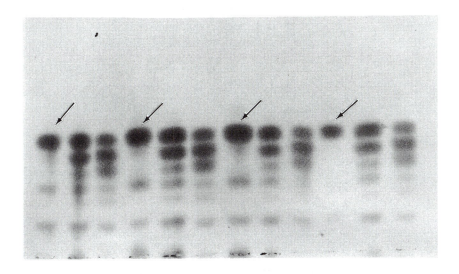

Fig. 4.10 Electrophoretic appearance of the rainbow trout serum enzyme malate dehydrogenase; a tetrameric molecule controlled by two loci, showing from left to right 4 groups of fish each comprising a double homozygote (arrowed), a fish with one slow allele and another with two slow alleles.

Despite the overwhelming importance of electrophoretic variation in the study of population genetics, to be discussed in Chapter 6, there are only a few examples of simple breeding programmes in which the nature of the inheritance of these characteristics is explored. The general assumption of genetic control in all the other cases depends upon extrapolations which will be described in Chapter 6.

Simple examples of genetic control of enzyme variation as revealed by electrophoresis in the rainbow trout are given by Thompson and Scott (1984). Phosphoglucomutase (Pgm) exhibited the simplest expression of a two-allele system in which each allele generated a distinctive band (Fig. 4.8) and the heterozygote both bands. In family studies the authors observed segregation ratios which were entirely consistent with Mendelian inheritance of two alleles at a single locus (Table 4.1). The alleles are given numbers which reflect the relative electrophoretic mobility of the enzymic forms with the commoner allele defined as 100. Pgm is a monomeric enzyme. A more complex situation was observed with superoxide dismutase (Sod), which is a dimeric molecule, i.e. comprising two different sub units, and for which three alleles were discerned. Single bands again characterized the homozygotes but the heterozgote expressed three-banded patterns reflecting each of the possible dimeric forms (Fig. 4.9). Table 4.2 shows the data for two of the alleles which, again, conformed to Mendelian principles.

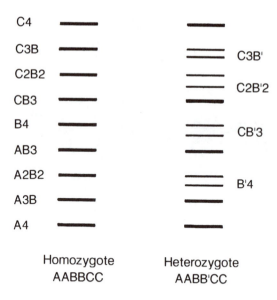

Homozygote   Heterozygote
AABBCC   AABB'CC

Fig. 4.11 Diagrammatic representation of the complex electrophoretic expression of alleles of the enzyme lactate dehydrogenase.

The most complex of the inheritance patterns reported by Thompson and Scott (1984) was for malate dehydrogenase (Mdh), another dimeric molecule but where each sub unit was under the control of a separate locus and both of the loci were represented by alternate alleles, five in number overall. Figure 4.10 also shows the patterns arising for Mdh but because of the common mobility of the molecule arising from the 100 allele at each of the two loci there is some confounding of genotype/phenotype identity. Thus a somewhat simpler overall pattern was observed than might otherwise be expected from the random assortment of four subunits to make a dimeric molecule. Simple consideration will show that 10 bands are possible, variously expressed by 9 genotypes. Table 4.2 shows the data of Thompson and Scott which are further simplified because of the absence of the rarer alleles at the two loci. This type of duplicated locus (Bailey *et al.*, 1970) is quite common in electrophoretic analysis and the question of its origin and inheritance will be further assessed in Chapter 6.

Similar studies by Utter *et al.* (1973) using rainbow trout and various Pacific salmon showed simple Mendelian inheritance for Pgm and Sod (also termed tetrazolium oxidase) and, in addition, the inheritance of a tetrameric enzyme lactate dehydrogenase (Ldh). Here too the data on cross matings conform to Mendelian principles even though the heterozygote banding patterns are very complex and difficult to interpret. The situation with Ldh is also complicated by the existence of multiple loci and differential expression in different tissues. Wright and Atherton (1970) describe five loci coding for Ldh in the brook trout, *Salvelinus fontinalis*, identified as A B C D and

E. The first two produce subunits in most tissues whilst C is expressed in the eye and the brain and D and E are expressed only in muscle tissue. Wright and Atherton describe the expression of three alleles at the B locus in eye tissue which conforms to a simple Mendelian pattern of inheritance. The diagnostic patterns for the B alleles are shown diagrammatically in Fig. 4.11. The heterotetrameres involving the A sub-unit are left out for clarity, but the confusion which could arise with multiple alleles at four loci, which is theoretically possible in muscle tissue, seems likely to defy analysis.

The inheritance of electrophoretic variation is thus basically simple but open to much complication due to the permutations of multilocus effects, multiple alleles and polymeric molecular structures. See also Chapter 6.

4.7 MISCELLANEOUS CHARACTERISTICS

Other than the obvious visual characteristics of colour, scale pattern and fin and body shape, there are few examples of simple Mendelian inheritance in fish. Some studies on blood groups will be described in the next chapter together with haemoglobin variants and the vast array of other proteins revealed by electrophoresis. The extent to which these are used positively in fish breeding, for example as breed tags, is limited but their involvement in population studies in natural and farmed species is wide and still has much potential for further development.

There are no examples of major histocompatibility complexes (MHC) in fish. These are the complex, highly polymorphic genes which exist in higher vertebrates and provide for the wide spectrum of antigen–antibody reactions observed in these animals. Fish do possess cellular immune defence mechanisms, and Kallmann (1964) has estimated the number of genes controlling the immune response which underlies the rejection of tissue transplants in fish. Similarly, Stet *et al.* (1990) describe a red blood cell antigenic complex in gynogenetic carp (Chapter 12) and suggest, in fact, that it might represent an example of fish MHC.

No major loci have been observed, nor is there any simple genetic determination for specific disease resistance. Major genes affecting tumour formation have been studied intensively in platies (Anders *et al.*, 1973). They mostly operate only in hybrids (*Xiphophorus helleri* × *X. maculatus*) and their genetic significance probably lies in the nature of their uncontrolled expression and its relevance to breakdown within cellular control mechanisms in hybrids. These are of interest to breeders, however, in that they often affect melanophore patterns and intensity of pigmentation and seem likely to be highly responsive to selection.

Growth rate in fish is of great importance to breeders for obvious reasons. No positive major gene effects have been observed here although some pleiotropic consequences to growth have been postulated for other gene-controlled processes. Simple genetic determination has been claimed for dwarf-type strains of fish used for fish farming where such attributes are an advantage. Such studies were briefly reviewed by Kirpichnikov (1981). It seems reasonable to assume that any major gene effect in fish will have some consequences for growth, food conversion efficiency, viability and sundry other 'commercial' or 'fitness' traits.

Chapter five

Quantitative genetics:
I. Metrical characteristics

5.1 INTRODUCTION

Quantitative genetics includes the study of two quite different branches of heredity. The first, the genetic control of metrical traits – things one can measure such as size, growth rate, survival, or numbers of scales or vertebrae – is of very great importance to the breeding of animal and plant species used by Man. The second, on the structure of populations, deals in allele frequencies and is also of value in breeding practice, but its greatest impact is on the study of evolutionary mechanisms in the broader sense: i.e. the dynamics of changes in the genetic structure of populations under selection, whether that selection is natural or artificial.

One aspect of the genetics of metrical traits, namely selective breeding to improve performance, is of paramount importance and, to many, the most significant of genetic approaches to animal and plant husbandry. We shall see that this branch of genetics attempts to follow the fate of a multiplicity of genes, each of which has only a small effect and is indistinguishable from all of the others. These minor genes are not definable in the way we have seen for the major genes which have obvious Mendelian hereditary pathways. This generalized multifactorial view is, perhaps, too simple but it has led to the terminology polygene (p. 66) and polygenic to describe the inherited units and their mode of inheritance. Polygenic inheritance is not measured in individuals but in populations, and it is the numbers of polygenes transmitted from one individual to the next which is of importance.

The second branch of quantitative genetics also concerns populations but this time with respect to the behaviour of populations themselves. Population genetics per se addresses evolutionary aspects of species, subspecies or races and the degree to which artificial populations diverge and interact with each other or with wild populations. Population genetics also embraces the problems of inbreeding in natural and artificial populations and the consequences of intraspecific hybridization and heterosis, of vital importance in plant breeding and some aspects of animal breeding. These aspects of genetics will be discussed in the coming chapters.

Much use is made of Mendelian inheritance in population genetics but, as observed in the previous chapter, it is the distribution statistics of the alleles within populations which are of importance, not the nature of the traits themselves, except in those cases where the allele in question is depicted as having an impact on fitness. Any sort of genetic attribute can be used in population genetics, but the most noteworthy advances in the past 30 years have come from studies with electrophoretically detected variations which have simple genetic determination.

5.2 POLYGENIC INHERITANCE

The term **polygene** was proposed by Mather (1941) to identify genes which each had only a small effect on the magnitude of a measurable character such as height or length, or numbers of vertebrae or, for insects, numbers of bristles or wing veins. The continuous nature of the variation between individuals for such characters was seemingly at odds with the non-blending principle implicit in Mendelian models of inheritance, but the postulation of multifactorial gene control provided the means whereby the inheritance of continuously variable characters could be embraced by Mendelian genetics.

It is surprising how few polygenes are needed to provide for a more or less continuously variable inherited trait. In Chapter 4 we considered the inheritance of plate morphs in *Gasterosteus aculeatus* and the hypothesis that it was controlled by two gene loci each with two alleles. If the range of plate numbers is 6 to 30 and if a further assumption is made that each dominant allele contributes a fixed number of plates to the series for any fish, then a fish with four recessives would have 6 plates, one with one dominant 12 plates, with two dominants 18 plates, three dominants 24 plates and four dominants would produce a fish with 30 plates. If three loci comprising six alleles were involved the steps would be even smaller at 4 scales per dominant allele on top of the basic 6. Thus by increasing the number of possible genes involved it is possible to move towards a continuous or in the case of scales, quasi-continuous variation.

For a single locus with two alleles, the genotypes present in a population will be the two homozygote types and the heterozygote. Representing the alleles as A and a, the genotypic possibilities are AA, aa and Aa or aA. If the frequency of one allele is p and that of the other q (p + q = 1) then in a random mating situation the union of alleles via fertilization is random and the probability of a fertilized egg having two of one sort of allele or the other is $p \times p$ or $q \times q$, i.e. p^2 and q^2, while the heterozygote is either $p \times q$ or $q \times p$, i.e. 2 pq. The frequencies of genotypes in a population is thus the expansion of:

$$(p + q)^2 = p^2 + 2pq + q^2$$

for genotypes AA, Aa, and aa, respectively.

If we take two gene loci, each with two alleles, then the possible genotypes are expressed as the products of a Latin square and their frequencies are

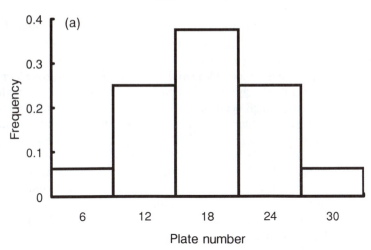

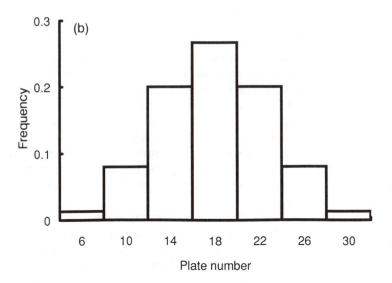

Fig. 5.1 Hypothetical frequency polygons for the genetic control of plate number in the stickleback: (a) two genes, each with two alleles; (b) three genes, each with two alleles.

likewise produced by the product $(p + q)^2(a + b)^2$. If we assume, for the hypothetical two-gene scale example, that each allele at each gene locus has a frequency of 0.5, we can work out the expected frequencies of the various levels of dominant alleles per genotype from the expansion of $(p + q)^4$. These frequencies are plotted as Fig. 5.1(a) and it can be seen that a symmetrical distribution is produced – it would be asymmetrical if allele frequencies in the hypothetical population were not all 0.5. Figure 5.1(a) depicts a coarsely graded distribution, but if three loci and two alleles each are postulated it becomes much finer (Fig. 5.1(b)) with the frequencies generated by the expansion of $(p + q)^6$; again, assuming that all alleles have frequencies of 0.5.

Thus with only a few gene loci with small additive effects it is possible not only to construct continuously variable characters but also to explain the frequency distributions which occur in natural and in reared populations of fish. It is believed that polygenes are much more numerous than in the hypothetical example used here. The effect of a large number of genes, plus the smoothing effect of environmental factors which have plus and minus effects on all measurable characteristics, leads to smooth variations and distributions. Metrical traits are therefore assumed to be determined by polygenic and environmental agencies working together. The task of quantitative genetics in this context is chiefly to estimate the extent of the genetic component of the variability.

5.3 HERITABILITY

A huge theoretical background to quantitative genetics has been built, admirably reviewed by Falconer (1981). It is beyond the scope of this book to do more than touch on some simple aspects. One of these is the assessment of the genetic and environmental control of measurable characteristics, such as performance traits, defined by the concept of **heritability.**

Most measurable characteristics in living things have an inverse bell-shaped frequency distribution which approximates to the mathematically defined 'normal distribution' in which the key parameters are the mean and the standard deviation, which is the square root of the variance (the variance being computed as the mean of the summed squared deviations of individual observations from the mean).

In statistics, the standard deviation is the normal parameter used in calculations: in genetics the variance is usually used to describe populations.

The value of the variance (V) can be apportioned to an environmental component (V_E) and a genetic component (V_G) and the latter can be further divided into additive (V_A) and interactive (V_I) components. The additive variance contributes most to the similarity between relatives and provides for a successful selection programme for genetic improvement. Heritability (h^2) is that part of total variance which is additive and genetic in nature. Thus:

$$V = V_E + V_I + V_A \qquad (5.1)$$

$$h^2 = V_A/V \qquad (5.2)$$

and for a selection programme for improvement,

$$R = h^2 S \qquad (5.3)$$

where S is the selection pressure measured as the difference between the mean of the whole population and the mean of that group of organisms chosen to be parents of the next generation, and R is the selection response expressed in similar terms, i.e. the difference between the mean of the original population and that of the next. This is a very simple model and it fits many observations with experimental organisms and cultivated animals and plants, but it has had a very controversial history within fish breeding, particularly with respect to that ever-popular theme, the improvement of growth rate. Part of the problem lies with the rigour required to estimate h^2 or to conduct a selection programme.

The estimation of h^2 is done either by measuring correlations between relatives, such as offspring and parents, or by recording progress under a selection programme – where h^2 is assessed as the slope of the improvement curve, and sometimes called the **realized heritability**. Full details of all the complexities involved in these estimations are given by Falconer (1981), but for fish it is sufficient to stress at this point that environmental factors, including maternal effects, are of extreme importance. In addition, control of environmental variance in fish farming is, at best, very difficult.

5.4 GROWTH RATE

This more than any other commercial trait has dominated the discussion on genetic improvement within fish farming. It is not so obviously important for the breeders of fancy fish. The reasons behind this dichotomy are probably that most fancy fish are small and have what is called a **determinate** growth pattern with a fixed final size. Food fish, on the other hand, seem to have what is termed an **indeterminate** growth pattern with growth continuing, albeit at an ever decreasing rate, during the lifetime of the fish. Various growth equations have been proposed to account for indeterminate growth (see Purdom, 1980) but the most significant feature of growth in farm fish is that achieved prior to sexual maturity. Most species of farmed fish are marketed before the onset of sexual maturity for a variety of reasons, but time and the debilitating consequences of the maturation process are the primary ones (Chapter 11). For fancy fish, time is not of great importance: if the fish do not all reach the required size by a given date they will do so at some later date – they are not slaughtered.

It is tempting to speculate that fish of a determinate growth pattern generate eggs (and spermatozoa) by direct metabolism of food whereas those of indeterminate growth rate use food for increasing body size and eventually convert body materials into sex products. It would follow that the latter would have advantage in continuous growth whereas the former would not.

Much confusion surrounds the economic importance of growth rate in fish farming. Dunham and Smitherman (1983), for example, state that "Even slight improvement in ... body weight would result in millions of kilograms of additional production". This would only be of significance if it was net gain, but increasing food requirements, pond space and other back-up facilities would be on the debit side and could be deployed anyway without the involvement of modest genetic gains. Most fish farm outputs at the gross level are limited by food, space, oxygen and temperature and almost never by genetic potential.

The main barrier to a practical application of quantitative genetics to growth rate, however, is the lack of strict control on the life cycle of the species used in fish farming. This is particularly true for species currently being developed as prospective farm fish – marine flatfish, for example – but even for the long-established species, carp and trout, and the more recent ones, such as salmon, the practical difficulties of culture technology make precision planning almost impossible. A secondary consequence of this is that improvements in husbandry are still easy to develop in fish farming and such developments year by year can be misinterpreted as gains from deliberate selective breeding, or as the unconscious development of domesticated status in the fish. Certainly one of the major selection programmes in trout culture (Donaldson and Olson, 1957), told in almost poetic terms more recently by Hines (1976), took place during the era of rapid development of trout farming technology, and the dramatic gains claimed on genetic interpretations must owe more to husbandry improvements than to genetics (Purdom, 1976). Donaldson began his programme with fish from a natural population and developed what has become the recognized Washington strain of rainbow trout. In their natural environment the fish attained a weight of about 600 g at sexual maturity after 4 years of growth. In the hatchery after a decade of selection, they were mature at 3 years of age at a weight of about 3 kg. One cannot compare natural and hatchery production, however, and with today's technology it would be possible to produce a 3 kg fish within 2 years of age just by good feeding. The Washington strain does not appear to have had much impact on commercial production even after introduction into the comparatively open market of United Kingdom trout production. The difficulties over the interpretation of this early attempt at selective improvement of fish growth rate could have been avoided by the maintenance of an unselected control line. This is an easy criticism to make in hindsight, however, and the resources that would have been needed to comply with it are equally obvious. This sort of study with fish of commercial importance requires a very heavy commitment of resources and is probably beyond all but the largest commercial concerns or the really committed government organization.

The most systematic study of the inheritance of growth rate in fish is that conducted by the Israeli carp breeding school under Moav and Wohlfarth during the 1950s and 1960s, (Moav and Wohlfarth, 1976). At a time when fish culture was offering the world a new source of high-quality protein foods, and genetics was tipped to play a substantial role in its development, the overall result of five generations of selection for improved growth rate was disappointingly negative. Moav and Wohlfarth performed mass selection on fish grown under more or less normal fish farming conditions at a range of fish farms centred on an experimental station at Dor in Israel.

The selection programme began in 1965 from a population of fry of mixed parentage from diverse farms within Israel. This use of mixed parents is a

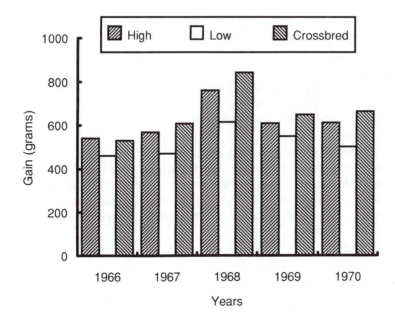

Fig. 5.2 Mean annual weight gain in carp selected for growth rate or maintained as crossbred controls (data from Moav and Wohlfarth, 1976).

standard approach undertaken to maximize genetic variance. Fry were selected for high or low growth rate at the outset, but after one generation the selection programme was based on weight after a growing season. Unselected controls were started – again a standard selection procedure – but failed after one generation. A crossbred control was then started and because fish can survive many years with a retained reproductive capacity, this stock provided a control to the end of the selection programme, 4 years later. Additional crossbred controls were produced in subsequent generations from the selected lines.

The results of the selection programme are summarized in Fig. 5.2. Selected lines and controls were maintained in ponds at the commercial farms and at the experimental station at Dor. The outcome, which must have disappointed the authors, was that no improvement occurred in the line selected for faster growth but that some deterioration possibly took place. In the line selected for low growth rate there was some evidence of reduced performance, i.e. a selection response. In a complex discussion, the authors concluded that failure to improve growth rate arose because the additive variance for this trait had reached a plateau due to natural selection and past, unconscious, artificial selection. This conclusion is not easy to reconcile with studies on experimental animals such as the fruit fly *Drosophila* and the mouse where the plateau condition takes many generations of careful selection for full expression, and response in the opposite direction at any time is rapid.

A simpler explanation of the lack of response for increased growth rate lies in the very great dependence of this trait on environmental conditions. Within the different fish farms used in the study, growth rate varied by as much as 300% within any group of fish for one year. Variation was also very great from year to year, with the weight increment of the standard crossbred showing over 200% variation from one year to another.

Weight is a difficult parameter with which to measure fish. In early growth phases, with constant temperature, the growth curve (the growth of weight against time) is often cubic – that is, it is described by a cubic equation and depicts a 'compound interest' increase (curved line in Fig. 5.3) – although as other factors such as sexual maturation intrude, the rate of increase in growth rate declines. Thus growth rate is dependent on size, and Moav and Wohlfarth were obliged to make corrections to the incremental weight gains to compensate for differences in the initial weight of fish. It is these corrected weight gains which provided the core data for the selection trials and are summarized in Fig. 5.2.

A lot less confusion would arise over fish growth rates if they were expressed in linear terms such as length or the cube root of weight. Figure 5.3 shows growth in weight and length of a fast-growing strain of rainbow trout fed as near to satiation as possible in a trough under a constant temperature regime.

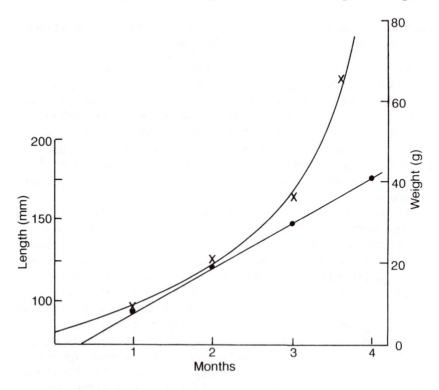

Fig. 5.3 Growth in terms of length • and weight (x) in a fast growing strain of rainbow trout, *Oncorhynchus mykiss.*

Growth rate in terms of increase in weight per unit of time is meaningless unless expressed as specific growth rate (proportional increase in weight), whereas the linearity of growth in length is obvious and the rate is simply the angle of the line to the abscissa and, for young fish at least, is independent of size. Once this 'pure' form of growth ceases, of course, the situation is different but in most fish culture systems the early growth is decisive.

In a selection programme on the popular North American farmed catfish species *Ictalurus punctatus*, Bondari (1983) also used a deliberately outbred population in an attempt to avoid any possible genetic plateau. Only one generation of selection was imposed but based on family selection instead of mass selection. Selection on the basis of family performance is generally thought to be more effective than mass selection when low values of h^2 are encountered.

Some response was generated, but the nature of it was unclear. Selection was performed during the early months of growth and selected parents were then raised to 5 years of age prior to spawning. At spawning, all crosses between fast-selected (W^+) and low-selected (W^-) were performed, and the results showed that $W^+ \times W^+$ and $W^+ \times W^-$ offspring performed equally well and better than offspring of $W^- \times W^+$ and $W^- \times W^-$. All selected groups did better than the control, however, so the quantitative picture is far from clear and from an experiment lasting over 5 years disappointingly inconclusive. In channel catfish, as in carp, environmental causes of growth rate variance are predominant (Klar, *et al.*, 1988).

Attempts to measure h^2 and hence provide the basis for predicting selection response have been very numerous, particularly in salmon and trout. Kinghorn (1983) reviewed many of these studies and recommended that "predictions ... should be interpreted with caution". Levels of h^2 for length or weight at a given age ranged from 0 to over 0.5 (h^2, of course, can vary from 0 to 1). Very little evidence has been advanced in support of the higher values which, if true, would imply that selective improvement of growth rate would be easy to achieve.

One major study where improvement in growth rate was achieved over a relatively short sequence of generations of selection in rainbow trout was reported by Kincaid *et al.* (1977). The programme entailed three generations of family selection as opposed to the mass selection normally chosen for selection programmes in fish. Family means of 2-year-old parents were measured for 43 F_1 families and 8 were selected for the next generation. When the same parents reached 3 years of age the process was repeated to give 13 additional families, selected from a total of 78 matings. Weight of the fish at 147 days of age was the selection criterion, timed from the date of fertilization of the eggs. Fish of the selected families were reared to 2 years of age for the initiation of the next generation of selection and for testing. Selected parents were retained for a further batch of egg production at 3 years of age but these offspring were used only for testing, not for perpetuating the selection programme.

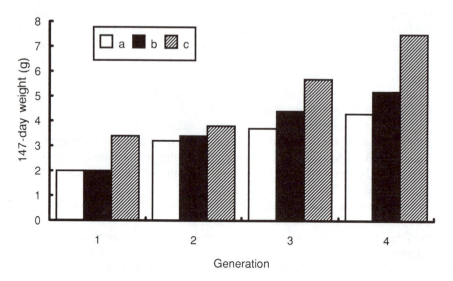

Fig. 5.4 Weight of rainbow trout at 147 days after hatching: open columns, unselected fish from 2-year-old parents; black columns, selected fish from 2-year-old parents; hatched columns, selected fish from 3-year-old parents.

Figure 5.4 shows the overall result indicating some progressive gain at each generation. There are, however, serious limitations to the methods employed in the selection programme and in the tests. Since the non-selected line improved dramatically over the period of the programme it is clear that very considerable general management improvements had taken place. Secondly, the manner of feeding the test populations – as a percentage of body weight each day – is a clear way in which bias can be introduced. The main problem, however, lies in the age of parents during the selection phase and the very obvious maternal effects on growth rate. The latter is clearly indicated by the superiority of the offspring from 3-year-old parents and can be explained simply on the basis of egg size. At 2 years of age, female rainbow trout are only just at the point of sexual maturity. At this time, the bigger fish will be expected to produce larger eggs for purely non-genetic reasons. Large eggs generate large fry and the advantage lasts well beyond 147 days. A similar confounding of genetic gains by maternal effects can be discerned in studies by Friars *et al.* (1990) on grilse length in Atlantic salmon. There are therefore real difficulties in the interpretation of these results but again, there is evidence of very large environmental variances and these are amenable to improvement by management practice which is no less valuable than genetic improvement and probably a good deal cheaper to achieve.

Other positive responses to selection in salmonids include studies by Gjedrem (1979) and Hershberger *et al.* (1990b), but in neither case were simple, unselected controls maintained and the evidence remains flimsy.

Other selection programmes for improvement of growth rates in fish have failed to generate a positive response. In tilapia, for example, mass selection failed even though parent/offspring correlations indicated reasonable heritabilities (Hulata *et al.*, 1986; Huang and Liao, 1990). Lack of response was also reported in tilapia by Teichert-Coddington (1988). An attempt to select for 56-day weight in a fish (*Gambusia affinis*) with a determinate growth pattern (Busack, 1983) seems to have ended disastrously.

The most telling criticism against the not inconsiderable amount of work which has been done on growth rate improvement in fish is that it does not appear to have contributed anything to commercial practice. Not only is there no clear evidence of farmers using specifically identified improved stocks, there is also good evidence that farmers are content to use stock which is known to be of lower growth rate than others. Much recent work on Atlantic salmon may be subject to 'commercial-in-confidence' restrictions because of the high business profile of the organizations taking part in these endeavours. Only time will tell whether or not suitable progress can be made in the commercial use of developed strains.

It may seem odd to discuss at great length the difficulties of interpreting growth rate selection programmes. So much effort has been put into studies of this sort, however, over the last half century and the simplistic views are so often paraded, even today, that it is necessary to caution against further wasted effort in this field. There are many ways of improving growth rate by environmental manipulation and by other genetic means – and there are traits more worthy for genetic selection than growth rate itself.

Fish in populations display hierarchical tendencies and the observation of burgeoning size ranges is common amongst fish breeders. In salmonids its impact is so great that frequent sorting for size is essential during the early months of growth of young fish. The science of hierarchy has been studied more widely, however, in other sorts of fish. In carp, a classical series of studies by Japanese authors (Nakamura and Kasahara, 1955, 1956, 1957, 1961) translated by Wohlfarth (1977) described the establishment of stable size hierarchies and the transient status of fish when transferred out of their natural population. Parallel attempts to quantify the consequences of hierarchy on size distribution were attempted by Purdom (1974) with marine flatfish. Using the coefficient of variance (the variance divided by the square of the mean) as the indicator of population stability, it was possible to demonstrate the development of hierarchies and the consequences of their elimination.

Hierarchical variation is essentially environmental. Attempts to select for size on such bases seem more likely to be selecting for aggression, and a consequence may be more variance, not more growth. A very elegant demonstration of the non-genetic origin of size variation in cyprinids came from the study of growth rate variation in ginbuna carp (Nakanishi and Onozato, 1987). These all-female forms of silver carp, *Carassius gibelio langsdorfii*, reproduce without meiosis and exist as natural clones, i.e. families with individuals which are genetically all of the same type (Chapter 10). When such genetically identical (isogenic) fish were reared they developed the same form of skewed size distribution, including the so called 'shoot' carp or 'jumpers' found from genetically mixed populations.

Under these circumstances, the shoot carp clearly have no genetic advantage over any other fish, but it is conceivable that in natural heterogeneous populations some genetic variance for aggressive feeding might predispose some individuals towards faster growth. Under such conditions selection might be expected to increase levels of aggression rather than growth rate overall. Durrante and Doyle (1990) discuss such concepts in relation to trials with *Oryzias latipes* and conclude that the model is feasible under conditions of forced social interaction, i.e. competition for food.

5.5 STOCKS AND STRAINS

Although the prospect of genetic gain from selection for increased growth rate seems slight compared with the effort required to maintain the selection programmes, there are obvious gains to be made in controlling environmental determinants of growth rate, and in time, these may reach a level of development which permits a more promising approach to genetic selection for growth. At present, environmental variance for growth rate seems far too large for effective genetic management per se. There is, however, one further genetic approach which entails choice rather than selection in its fullest genetic sense. In nature a fish species may often occupy quite diverse habitats and the concept of natural stock distinctions is quite common. Under varying conditions of growth it might be assumed that in evolutionary time scales fish have become adapted to their environments to make the most effective use of feeding potential. Thus there might be little point in retaining a large growth potential under poor growth conditions and, vice versa, where opportunities for growth occur regularly or even sporadically it would be advantageous to be able to utilize them. Thus stocks within a species, e.g. geographic 'races', might be expected to show variation for growth rate and other characteristics – developed over tens of thousands of years or longer. Recent domestication may have added to this by the natural choice of better 'strains'. There are no generally accepted definitions of the terms stock, strain, line or variety. They each represent genetic derivations from 'species' with some natural and some domestication implications.

Natural and semidomesticated strains of fish with different growth rates do exist; choice of the more suitable types is obviously a simple and cheap way of achieving the sort of progress which in a selection programme would be uncertain to say the least and very expensive.

Clear differences between strains emerged from the Israeli programme on the quantitative genetics of common carp (p. 70). Wohlfarth *et al.* (1961) describe 'social' differences but conclude that in practice pond production rates were determined by genetic–environment interaction and it mattered little what 'race' was used. Later studies with simple genetic scale pattern types

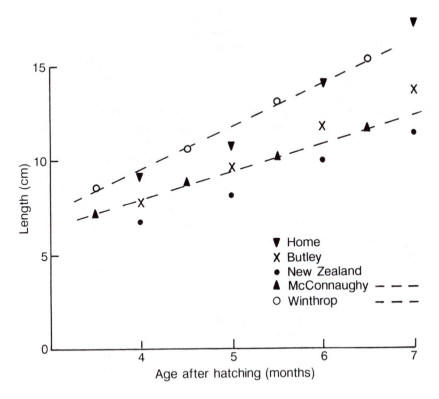

Fig. 5.5 Growth in length of rainbow trout from five individual domestic strains held under standard conditions. Regression lines for the best and the worst performers are shown.

(Chapter 2) suggested that the scaly form grew best, followed in order of performance by mirror, leather and line (Wohlfarth *et al.*, 1963). The poorer performance of the non-scaly variety seems not to have been any barrier to their use in farms and fisheries world-wide! More significant strain differences were established by Moav *et al.* (1964) in a comparison of Israeli carp with lines imported from Europe, and by Moav *et al.* (1975) in a comparison of Asian and European races of the common carp.

A similar analysis of strains of rainbow trout from different regions has also demonstrated very large differences in growth rate. The Ministry of Agriculture Fisheries and Food in England and Wales established a collection of strains of these fish chosen primarily for the season at which they assumed sexual maturity in a programme designed to reduce UK dependence on imported eggs. The spawning characteristics will be described later but an additional finding was that growth rates varied widely. Figure 5.5 illustrates the growth rate of immature fish of five strains, the best of which grew in length 50% more rapidly than the worst. Differences in weight after a period of growth were even more dramatic and attempts to set up coded replicate rearing tanks, unspecified to the husbandryman, were thwarted by the obvious differences in feeding rates. It is noteworthy that all of the imported domesticated strains were important farm stocks in their own right!

The growth rate differences between these rainbow trout strains were genetic in that successive generations exhibited a continuity of the trait and, more specifically, hybrids between slow and fast growers grew at an intermediate rate. The question remains whether such strain differences represent genetically derived improvements. To assess this, hybrids, F_2 and backcross generations were compared with straight crosses under controlled growth conditions with specific attention to growth rate variances. Classical genetic models predict that variances in parental and F_1 generations for two different parental types remain similar but that in the F_2 or backcross, an increase in variance can be expected because of the reassortment of alleles within the developing gametes of the F_1 hybrid. Since variances and means are correlated in normal or near-normal distributions, the assessment was made on the basis of coefficients of variation, the ratio of the standard deviation to the mean expressed in percentage terms. The results for length and weight are shown in Figs 5.6 and 5.7. Clearly, the coefficients increased with age (environmental) and there was more variance in the straight crosses than in the backcross or F_2. The uniformity of the increase in the coefficient with age across all genotypes is evidence of a uniform environmental effect on variance. The comparisons on absolute levels, however, are contrary to the classical genetic model. Thus the huge differences in growth rate between the strains appear to be genetic, but not additively genetic. Regrettably, resource limitations

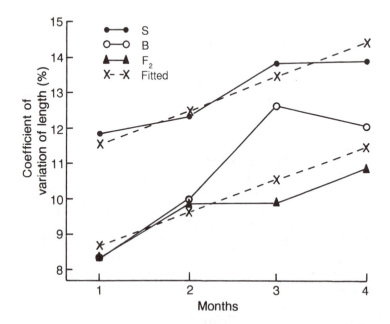

Fig. 5.6 Coefficients of variation for length for three categories of genotypes (straight crosses (S), backcrosses (B) and F_2 crosses) in monthly samples of rainbow trout.

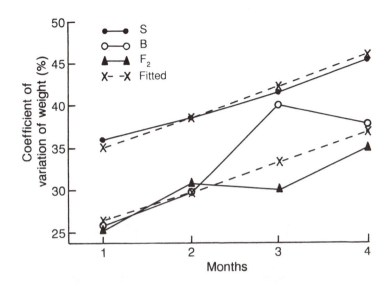

Fig. 5.7 Coefficients of variation for weight in the samples described in Fig. 5.6.

precluded the final test of this lack of additivity, i.e. a selection programme started from the F_2.

Strain differences for growth rate have been described for Atlantic salmon and it seems probable that any species of fish for which discrete natural populations exist should show some variation of this type and be fruitful sources of stock for growth rate assessment and use in farming.

5.6 SPAWNING TIME

The full domestication of an animal destined for intensive cultivation cannot be achieved until the seasonal limitations of spawning have been removed so as to permit breeding all the year round. In fish, as in many organisms, the onset of sexual reproduction is seasonal and is determined by climatic conditions such as temperature and daylength. Temperature appears to be the important cue for cyprinid fishes but light, in the form of daylength, controls the processes of sexual maturation and spawning in salmonids and marine flatfish. This sort of control probably exists in most fish species native to temperature zones. On the other hand, in the tropics where summer and winter do not offer quite such different prospects for survival and growth, no such constraints need apply.

In salmonids and flatfish, spawning can be controlled by manipulating daylength (Scott *et al.*, 1984) so as to generate all-year-round egg supplies, but the methodology is expensive and not always reliable. Similarly, hormonal

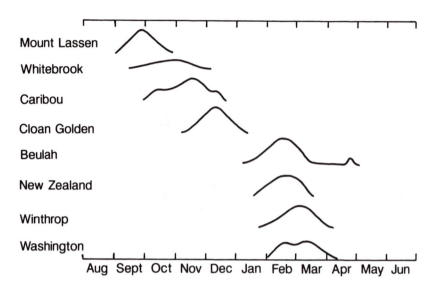

Fig. 5.8 The spread of spawning times in selected domesticated strains of rainbow trout.

control of spawning is possible (Hunter and Donaldson, 1983), but only to a small degree and again not always reliably. For an established fish cultivation industry the use of individual strains at different times of the year is the best solution. This seems practical only for rainbow trout at present. The domesticated strains of this species in use world-wide have a range of spawning times which spans the entire year. Figure 5.8 summarizes the spawning times of the various strains which were collected under government auspices for use in the UK. It can be seen that the autumn, winter and spring periods were adequately provided for. In addition, one strain (not shown) spawned in mid-summer; this was special, in that it spawned twice a year but, unfortunately, its other commercial characteristics were not good and commercial usage was not favoured.

There is little doubt that spawning time in rainbow trout is genetically determined and hybrids between the extremes in Fig. 5.8, facilitated by sperm cryopreservation, were intermediate between the two parents. There is no further knowledge of the way in which genetic control is exerted nor on the origin of the various stocks, but anecdotal evidence suggests that the wide range of spawning times has been developed by selection and that a shift of about 1 month can be achieved in four or five generations of selection. There is no reason to doubt that similar selection programmes on other trout or salmon species would generate progress. A more disturbing prospect is that the seasonal management of commercial fisheries could have a selective effect in this context.

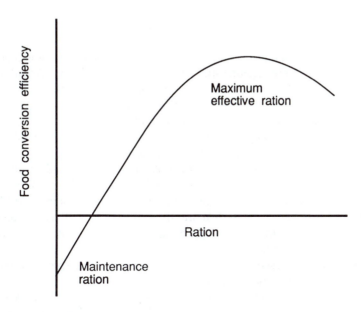

Fig. 5.9 Diagrammatic representation of the relationship between feeding rate and food conversion efficiency. See text for interpretation.

5.7 FOOD CONVERSION EFFICIENCY

Artificial food is provided in all intensive and most extensive fish farming systems. The diets are very expensive when they have a high animal protein content, and efficiency of use should be a high priority. The nature of food conversion efficiency is such, however, that direct genetic improvement seems no more likely to succeed than selection for faster growth rate.

Food conversion efficiency (FCE) is defined as the ratio of weight gain to food given, irrespective of the composition of diet or fish flesh. The general relationship between the level of diet and FCE is shown in Fig. 5.9. At low dietary levels the fish lose weight and FCE is negative. At a given level of diet, often around 0.1% of body weight, but dependent on temperature and fish size, the fish neither gain nor lose weight and this level of diet is defined as the maintenance ration. Above this level, increasing the rations produces an increased weight increment up to a maximum beyond which further addition to the diet has no effect. In consequence, FCE starts below zero, rises with increasing rations to a maximum and then declines. There are several points on these curves where selection might be applied, but in practice the parameters are so dependent on uncontrollable environmental variation that quantitative genetic management seems impractical.

5.8 DISEASE RESISTANCE

Disease is a highly significant problem in all forms of fish culture and includes invasion by infectious agents such as viruses, bacteria, fungi and some protozoa, attack from a variety of metazoan pests, and susceptibility to inborn errors of metabolism. Examples of the last such as crooked spines and abnormal swim bladders are clearly genetic and were discussed in Chapter 3. The genetic control, if any, of responses to invasion by other organisms is, however, not so clear cut and not very well documented.

Fish immune response systems are not so well developed as those of mammals. Two basic systems are humoral (i.e. antibody mediated) and cellular (i.e. under the control of major histocompatibility systems). Antibody formation is only very weak in fish and the cellular types of defence mechanisms seem to be the more important components of immunity. For reviews see Manning *et al.* (1987).

Several more or less specious claims for disease resistance to specific pathogens in farm fish have been made but there is little real evidence for a quantitative genetic control of such characters (Chevassus and Dorson, 1990). Detecting phenotypic variation for such traits is confounded by natural immunity anyway, and although strains of carp and trout have been shown to produce different responses to infection, the genetic nature of this variability is not established.

Attempts to improve by selection the resistance of fish to specific diseases do not appear to have been highly successful. Ehlinger (1964, 1977) selected brook trout, *Salvelinus fontinalis*, and brown trout, *Salmo trutta*, for resistance to furunculosis, an endemic disease produced by a bacterium, *Aeromonas salmonicida*. Selection methods were crude over the first 10 years. A range of geographic hatchery strains was assembled and exposed to the disease in the summer by cohabitation of potential broodstock with known infected fish. The strains with better survival were perpetuated either individually or by hybridization with other good performers. The 'selected' parents were those fish which survived the disease challenge, and no quantitative data were presented for possible analysis of heritability of the character.

After 10 years and presumably three or four generations, the offspring of individual groups were tested for their response to a laboratory challenge in the form of cultured bacteria. For brook trout the data on survival time after challenge showed that the unselected control did die more quickly than the average, 124 h after exposure versus 135 h. The brown trout data suggested that the survival rate of the unselected group was much lower (17%) than that of the average (44%). Thus positive 'gains' were recorded but impossible to quantify.

In field trials the 'selected' fish were placed in commercial farms and anecdotal-type evidence of benefit was given. The picture was confused, however, by the action of farm managers to control the disease outbreaks with antibiotics! Such are the problems of research under applied conditions.

In the second of his reports Ehlinger (1977) described a further 10-year programme. Selection for this period was based on the laboratory tests but included all sorts of other goals such as size of broodstock, body conformation and colour. The programme was beset with practical problems unrelated to the selection work per se such as flooding and diseases other than furunculosis. The results suggested that no further gains in resistance were recorded in brook trout but that in brown trout resistance was developed further in relation to the laboratory test. Field tests were disappointing, however, and the project was terminated in 1971. However, the developed strains were placed in fish hatcheries and Hulbert (1985) reports a comparison of one of them with an unselected strain of brown trout in a sport fishery administered by the New York State Department of Environmental Conservation. The 'furunculosis resistant' strain did survive better than the other in terms of anglers' catches and electrofishing surveys, but no conclusions were reached on disease implications and the two stocks were raised at different hatcheries, obviating any genetic analysis. This furunculosis programme is probably one of the best-recorded attempts to selectively improve a specific characteristic in trout and its disappointing outcome – furunculosis remains one of the most serious diseases of farmed and wild salmonids – is a reflection of the enormous difficulties posed by these traditional applied genetic approaches. The modern genetics of gene transfer may offer more direct pathways to these and other important goals and will be considered in Chapter 13.

5.9 SELECTIVE IMPROVEMENT IN FANCY FISH

Mixed success must be the most optimistic assessment of the achievements of quantitative genetics and selective improvement for those traits of commercial food fish which are of economic importance. The success of breeders of fancy fish of freshwater types, however, is undeniably spectacular and obvious for a variety of species commonly seen in ornamental pools and aquaria and in aquarist shops.

A contrast can be made between two styles of fish breeder. First, there are those who seek to breed wild species in captivity, often solving quite taxing problems in the process but aiming only to produce fish that resemble their wild relatives. The second style embraces only a handful of fish species and here the devotees seek to develop artificial strains which in their colour and shape deviate dramatically from anything in the wild. A further dichotomy in the latter style is the development of colour variants on the one hand and fancy strains with developed fin and body shapes on the other – the two are often combined but entail different approaches. The goldfish is the most diversely developed species (Fig. 5.10), closely followed by **koi** (nishikigoi) the ornamental common carp. Warm water fish such as members of the highly variable platy family, the mollies and the humble Guppy, have been selectively developed into bizarrely beautiful forms. Why this contrast between the food fish and the fancy fish? There are probably many reasons but three obvious ones are: first, the relative unimportance of environment in the determination of the fancy fish traits; secondly, the implications of major genes in their genetic control; and thirdly, the fact that observation alone is adequate for the practical application of selection – time-consuming, and awkward to manage, measurements are unnecessary. The development of the colour types such as the reds and blues in the Guppy and the kaleidoscope of varieties in the Siamese fighting fish, *Betta splendens*, (Wallbrunn, 1957), are achieved by selection or choice of specific genotypes, consciously or not, but the depth or vividness of the colour is often selected in the time-honoured fashion of amateur geneticists seeking gradual gains. None of this selective improvement is scientifically catalogued but its effect is clearly discernible in that small range of fish species for which specialist hobbies exist. The results can be seen in aquarists shows and are beautifully illustrated in the popular monthly magazines such as the *Tropical Fish Hobbyist* and *Aquarist and Pondkeeper*.

Major gene influence is also manifest in many of the characteristics of fancy fish. Colour has already been mentioned but alleles for body shape and fin length are also implicated. Thus the hi-fin, lyre-tail and brush-tail varieties of the platy can be traced to major gene effects, possibly enhanced by hybridization.

To take environmental factors first, the body and fin shapes which characterize most fancy forms of fish owe little to environmental variables. In most natural fish species the morphometric indices which specify body shape and fin conformation are remarkably stable compared with the characteristics of growth, food conversion, fecundity etc., so much so that taxonomy is heavily reliant on such characteristics. Thus any deviation is likely to have a significant genetic element and respond naturally and easily to unsophisticated selection. The flowing fins, protruding eyes and deformed spines of the goldfish fancy types are not expressions of the natural range of variation in the species.

For colour, the situation is slightly different. We have seen in Chapter 3 that colour variants often depend upon the expression of simple Mendelian major genes. However, the number of possible phenotypes involving epistatic interactions and dominance is such that the variety can be bewildering and beyond the analytical scope of normal genetic explanation. This complexity is augmented by the presence of modifier genes enhanced by selection. Furthermore, the flowing tails and caudal fins of the veil-tail and delta-tail varieties of the Guppy almost certainly also arise from the combined effect of major genes and selective improvement. It is highly probable that the phenotypic consequences of a major gene mutant allele are so abundantly pleiotropic and interactive with other genes in the genotype that they are immediately responsible for additive genetic variance – in other words, modifiers will come into play whenever the normal phenotype is significantly disturbed. This genetic model does not seem to have been tested. Fancy fish offer an obvious area in which to put this right.

Observation of exceptions must be an easier route to selective improvement than the careful measurement, selection and retention practices required for mass selection, let alone the complex handling procedures needed for family selection. This probably explains the observation that whatever performance traits distinguish our domesticated animal breeds, there can be no doubting the external distinguishing features. Compare, for example, the Aberdeen Angus beef breed with the Charolais, or the Rhode Island Red hen with the White Leghorn. So it is with fish, too.

Chapter six

Quantitative genetics: II. Population studies

6.1 INTRODUCTION

Up until about 25 years ago the study of population genetics was largely dominated by two concepts: first, the existence of polygenes with small effects of plus or minus value on metrical characteristics as discussed in Chapter 5 and, secondly, the existence of **mutations** – changes to the normal gene which produce more or less large, easily identifiable effects on the wild-type phenotype. Such changes are generally detrimental to fitness and are eliminated from populations such that mutant genes are rare occurrences and mutant phenotypes even rarer events in populations. The mutant genes themselves were kept at equilibrium level in populations by the opposite effects of natural selection and mutation. Mutation rates per gene locus are extremely low, at something less than 10^{-5} per generation. The frequencies of the mutant phenotypes were either proportional to mutation rates for dominant genes or equal to the square of the mutant gene frequency if recessive in nature. It is worth noting that mutant gene and allele are synonymous but the former often has connotations of undesirability.

Thus individuals within populations were deemed largely to be of uniform phenotype, and hence genotype, with genetic variation predominantly in the form of neutral polygenes or disadvantageous mutant major genes. Some exceptions were observed such as eye colour in Man, shell colour in snails and melanin pigmentation in moths but these polymorphisms were deemed to be special. The advent of biochemical genetics in the 1960s was to change this view of population genetics quite dramatically, but the seeds of change were sown long before this with the discovery of the human ABO blood group systems by Landsteiner at the turn of the century, at about the same time as the re-emergence of Mendel's work.

Blood groups or other manifestations of the immune system were the first biochemical genetic characteristics to be consciously assessed in connection with population studies in fish but they were quickly followed, and indeed largely superseded, by the development of electrophoretic methods for analysing variation at the protein level. The most modern developments now involve the analysis of DNA structure by the methods of molecular biology, in terms of the nature of mitochondrial DNA, repeat sequence DNA of the chromosomes or unique DNA sequences of individual genes themselves. These methods in fish will be briefly described, followed by a review of their application in fish breeding and population studies.

6.2 BLOOD GROUPS

The analysis of blood groups in animals traditionally involved blood typing by the observation of the reaction of red blood cells to soluble components of blood serum – the fluid left behind after blood has clotted. The reaction takes place by the linking of chemical moieties on the surface of the red blood cells, the **antigens**, to the soluble **antibodies** of the serum. The antibodies have more than one reaction site so that the reaction produces clumping of the red blood cells (agglutination) or, less commonly, the destruction of the cell membranes and release of haemoglobin and other cellular contents (lysis). It is the nature of the antigen that is normally under study, but some naturally occurring antibodies or agglutinins also reflect genetic characteristics. Blood groups were first discovered in Man in relation to blood transfusion practices and disease manifestation in new-born children, but now have wider relevance in general hereditary terms, and it is in this latter context that they have been studied in fish and other animals.

Four quite different antigen/antibody relationships can be defined (Fig. 6.1). The simplest is where the antigen and antibody occur naturally in a reciprocal way. The best-known example of this is the human ABO systems where naturally occurring antigen A in the cells is accompanied by antibody anti-B in the serum and where antigen B occurs on the cell walls and anti-A is found in the serum such that agglutination of A cells is achieved by B serum and vice versa. The dual phenotype AB has both sorts of antigen on the cell surface but produces no antibody; the O blood group has no antigens on the red blood cells but produces both antibodies in the serum. The phenotypes are produced by three alleles at a single locus – namely A, B and O – with genotypes AA and AO, BB and BO, AB and OO corresponding to blood groups (phenotypes) A, B, AB and O, respectively. Some of the earlier blood groups discovered in fish were similar to the human ABO system. Thus in the brown bullhead, *Ictalurus nebulosus*, natural agglutinins in the serum of individuals reacted with red blood cells from other individuals and there was a reciprocal pattern in which fish with either of the two antigen types of the red blood cells had the antibody of the other in the serum (Cushing and Durall, 1957). There did not appear to be a null allele corresponding to O in the human ABO blood group systems but a situation of this sort was reported in the big eye tuna, *Parathunnus sibi*, by Sprague *et al.* (1963).

The second and third types of relationships are where the antigens occur naturally but the corresponding antibodies must be made by exposing other animals to the antigens. Thus the use of the same species in which to induce

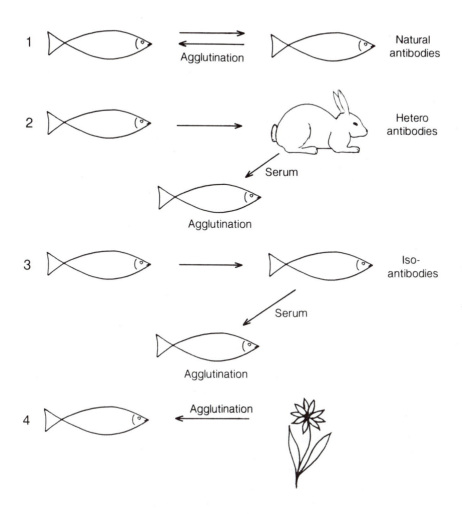

1 — Agglutination — Natural antibodies

2 — Serum — Hetero antibodies — Agglutination

3 — Serum — Iso-antibodies — Agglutination

4 — Agglutination

Fig. 6.1 Blood agglutination techniques used to define red cell antigens.

the production of antibodies generates isoagglutination procedures whilst the use of other animals, traditionally rabbits, generates heteroagglutination. The latter is much the more common procedure simply because fish are not very reactive in humoral immunological terms whilst rabbits, in common with other mammals, have very sophisticated immune response mechanisms. Nevertheless isoagglutinins have been described in some species of fish. For example, in rainbow trout, Ridgeway (1966) developed immune sera which were diagnostic for rainbow trout and also for other salmonids.

The heteroagglutinin examples in tuna described by Sprague *et al.* (1963) were likened to the situation in Man. However, the tuna blood groups were identified by using antibodies produced in rabbits or in cattle by injection of red blood cells from the fish. By absorbing the antisera with red blood cells from individual fish before testing others, the system was made specific: it should be noted, however, that the tuna ABO system does not really parallel the human system in which natural soluble antibodies exist complementary to the red blood cell surface antigens. The need for specially prepared reagents to detect the various fish blood groups is one of the reasons why these techniques have been superseded by electrophoresis; the reagents were not standardized.

The fourth type of blood typing involves the use of naturally occurring haemagglutinins where there is no relationship between the antigen holder or antibody donor. In other words the circumstances are accidental and have no biological significance other than in blood typing itself. The most widely known examples are of plant materials which contain haemagglutinins; a review of their use in fish is given by Utter *et al.* (1964).

In addition, some fish species have naturally occurring agglutinins in the serum which are reactive to specific red cell antigens in other species. Cumming (1967) lists serological cross-matching of 33 species of marine fish, finds natural agglutinins in many of them, and gives some detail of genetic differences in haddock, *Melanogrammus aeglifinus*, revealed by the use of sera from winter flounder, *Pseudopleuronectes americana*.

Application of blood group data

The use of blood group characteristics in the study of subpopulations of fish has been reviewed by DeLigny (1969) and Kirpichnikov (1981). A wide range of species and species groups has been studied and much individual and population variance has been described.

Individual variation in albacore, *Thunnus alalunga*, at a locus labelled Tg by Suzuki *et al.* (1959) was one of the earliest of such studies; interpretation of the data by Marr and Sprague (1963) supported the validity of the genetic status of the variation, and further suggested that a distinction between Pacific Ocean and Indian Ocean populations could be discerned. Such a conclusion is, perhaps, not surprising but an additional discovery was that while the data from the Pacific fish were compatible with the existence of one interbreeding population there, the data from the Indian Ocean suggested that the fish there comprised more than more population (p. 94). The various tuna species are impressive by their huge ocean migrations. They, perhaps more than any other group of fish, have been extensively studied by the use of blood groups (Fujino and Kang, 1968).

Distinction between geographically widely separated fish populations has also been demonstrated for other fish, e.g. herrings, *Clupea harengus*, and anchovy, *Engraulis encrasicolus*, in Soviet waters (Altukhov *et al.*, 1969).

Geographically more subtle distinctions have been observed for cod, *Gadus morhua*, where fish caught off Lofoten, which could be classified as coastal or Arctic by the shape of the otoliths (ear bones), also showed significantly different frequencies of a range of blood groups identified by the use of human typing sera (Moller, 1968).

Despite the initial success of blood-typing work, it has been superseded by the more practical and informative analysis of soluble proteins by electrophoresis. The main drawback to blood typing lies in the use of heteroimmune sera which are difficult to standardize. Isoimmune sera techniques have been developed for salmonids and have shown highly complex population structures (Ridgeway and Utter, 1964). With the large growth in fish cultivation practice since the 1960s there is probably scope for further development of natural or induced isoimmune techniques for the classification of fish breeds.

6.3 ELECTROPHORESIS OF SOLUBLE PROTEINS

By far the most popular way to study population genetics in fish has been by way of the electrophoretic analysis of soluble proteins. The simple mode of inheritance of this sort of molecular variation was described in Chapter 4 but breeding tests to define the genetic nature of the variations for most of the specific proteins that have been studied in fish have not been attempted. This is probably due to the fact that much of the earlier interest in fish population genetics centred on species that were commercially exploited in fisheries, and the technology to breed them in captivity simply did not exist. The large thrust in fish farming activity over the past decade or so is changing this and also introducing more ways of applying the principles of population genetics.

A general introduction to biochemical systematics is given by Ferguson (1980): what follows is only a brief outline of the subject.

Genetic complexities

As explained earlier, the basic genetics of electrophoretic variation is very simple because of the lack of any recessive/dominance relationships. The co-dominance of different alleles ensures that the pattern of bands, the phenotype, fully reflects the genotype. There is, however, one genetic complication and a host of biochemical or developmental complexities.

The genetic anomaly is the existence of **null alleles**; that is, those alleles which fail to produce any detectable protein. Without breeding tests, they are very difficult to identify and are postulated to explain observed genotypic frequencies. They ought not to be viable in the homozygous state, but crosses to test this do not appear to have been conducted.

Biochemical complexities

Biochemical complexities arise because of the structure of proteins. The simplest system is where the protein is a single molecule arising as a direct product of a gene. This monomeric structure generates a phenotype in which each allele is represented as a single band in the homozygote and as two bands in the heterozygote (as shown in Fig. 6.2(a)).

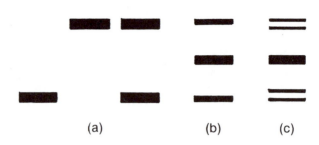

(a)   (b)   (c)

Fig. 6.2 Simplified representation of the electrophoretic phenotypes of monomeric, dimeric and tetrameric molecules in homozygous and heterozygous states. See text for interpretation.

Proteins also comprise complex molecules, polymers, in which two or more subunits produced by individual genes combine together. Where the protein is built of two molecules, a homozygote again shows the single band phenotype but the heterozygote shows three bands (as in Fig. 6.2(b)), by random combination of the two different molecules. If the molecule produced by one allele is A and that by the other is B, the heterozygote protein molecules will be AA, AB and BB in the ratio 1:2:1 just as for random assortment of genes (Chapter 2), but in this case involving intracellular proteins, not genes. For these dimeric proteins it is important to note that the two units of the finished protein are produced by the one locus. For more complex molecules, such as the tetrameres with four subunits, the options are wider because the subunits themselves may derive from one or two loci, depending upon the protein. The most straightforward cases are those where only one locus is involved. If we again use A and B as depicting the subunits produced by each of the two different alleles, the finished molecule may be of five types, A4, A3B, A2B2, AB3 and B4. These would be produced in the ratio 1:4:6:4:1 under random association but stereochemistry may influence the relative frequencies. Nevertheless, five-banded phenotypes indicate the existence of a tetrameric protein for which subunits are coded at a single locus and these are shown diagrammatically in Fig. 6.2(c).

Where two loci code for subunits of a tetrameric protein, highly complex phenotypes are possible with patterns involving several bands. The classic example here is the much-studied haemoglobin macromolecule which is built of four subunits, each with an α or β chain of amino acids determined, respectively, by genes on two different chromosomes (Dickerson and Geis, 1983). Defects in the chains or substitution of amino acids can lead to much phenotypic variability.

The different forms of proteins studied by electrophoresis include complexes such as haemoglobin, transferrin and some general proteins, but the overwhelming majority of soluble proteins studied in this way are enzymes, the catalysts of biochemical processes. Where different alleles at a single locus generate electrophoretically distinguishable enzymes, the separate entities are usually called **allozymes**. Where the different forms of enzyme independently derive from different loci, they are termed **isozymes** but there is considerable laxity in the use of these two terms.

We have discussed the allozyme phenotypes which arise from allelic variation. Genic variation can produce multiple isozymes which are not interactive but do appear on the same gels as each other and stain by the same diagnostic techniques. There is great variation between enzymes for isozyme number but in fish the esterases are very complex and so too are the lactate dehydrogenases. As in many other instances, the complexities arise out of so-called duplicate loci, the presence of gene pairs on two different chromosomes arising as an evolutionary consequence of doubling of the entire chromosome set (Chapter 12).

Developmental complexities

Enzymes have specific functions and often appear only in certain tissues. Where there are numerous isozymes, it is possible to misinterpret data as genetic variation when the variation is really developmental and mediated through the differential expression of different isozymes in different tissues. In fish, esterases often fall into this category and great caution is required in the interpretation of electrophoretic data.

Further developmental confusion can arise through the sequential change of isozyme content during the lifetime of individuals. The best example of this in fish concerns haemoglobins rather than enzymes. Much work has been done on haemoglobins in salmon and trout (Wilkins, 1972). In these fish, different haemoglobin types occur in the embryo, juvenile and adult stages of the life cycle and there is a gradual change from one form to the next. It therefore follows that great care must be exercised to avoid misinterpretation of these ontogenic successions.

Table 6.1 lists some of the enzymes studied by electrophoresis in fish including isozyme or allozyme characteristics and tissue specificity. International agreement has been reached on nomenclature and classification of isozymes and corresponding EC (Enzyme Council) numbers are given for the examples in Table 6.1.

Determining genetic models

In the absence of breeding data the postulation of genetic status for variation in population data can proceed stepwise.

The first step is to ensure that the variation is stable and repeatable and not an artefact of the electrophoretic or sampling techniques. For example, haemoglobin variation is dependent upon fish age, so one individual can express the variation at different times in its life history. Each haemoglobin type is itself genetically determined but by different genes acting at different times.

A second step is to assess the banding patterns in relation to the known or postulated subunit structure of the protein molecule. Thus for a monomeric

Table 6.1 Polymorphic isozymes commonly used in fish population studies with details of Enzyme Council number, number of alleles and illustrative references

E.C. No.	Enzyme	Locus	No. Alleles	Species	Reference
2.6.1.1.	Aspartate amino transferase	Aat-3	2	*Salmo salar*	Ryman, (1983)
1.1.1.40	Malic enzyme	Me-2	2	*Salmo salar*	Ryman, (1983)
1.1.1.14	Sorbitol dehydrogenase	Sdh-1	2	*Salmo salar*	Ryman, (1983)
1.1.1.27	Lactate dehydrogenase	Ldh-1	2	*Salmo trutta*	Ferguson and Fleming (1983)
1.1.1.27	Lactate dehydrogenase	Ldh-5	2	*Salmo trutta*	Ferguson and Fleming (1983)
2.7.3.2	Creatine kinase	Ck	2	*Salmo trutta*	Ferguson and Fleming (1983)
1.1.1.42	Isocitrate dehydrog- enase	Idh-1	4	*Oncorhyncus mykiss*	Allendorf, (1975)
1.15.1.1	Tetrazolium oxidase	To	3	*Oncorhyncus mykiss*	Allendorf, (1975)
1.1.1.8	Glycerophosphate dehydrogenase	αGpdh-1	2	*Oncorhyncus mykiss*	Allendorf, (1975)
1.1.1.27	Lactate dehydrogenase	Ldh-4	2	*Oncorhyncus kisutch*	Utter et al. (1980)
1.1.1.27	Lactate dehydrogenase	Ldh-C	2	*Cyprinus carpio*	Brody et al. (1979)
3.1.1.2	Esterase	Est-3	3	*Cyprinus carpio*	Brody et al. (1979)
2.7.5.1	Phosphoglucomutase	Pgm-B	3	*Cyprinus carpio*	Brody et al. (1979)
1.15.1.1	Superoxide dismutase	Sod-A		*Micropterus salmoides*	Phillip et al. (1983)
1.1.1.37	Malate dehydrogenase	Mdh-A	3	*Micropterus salmoides*	Phillip et al. (1983)
1.1.1.42	Isocitrate dehydroge- nase	Idh-B	2	*Micropterus salmoides*	Phillip et al. (1983)
1.1.1.44	6-Phosphogluconate dehydrogenase	6-Pgdh	2	*Oryzias latipes*	Saikazumi et al. (1983)

model the simple single-banded patterns should prevail, whereas for a dimeric molecule, three-banded and single-banded patterns should be seen.

Thirdly, the genetic model must be met, i.e. patterns must conform to the principle that only two allelic expressions at most can occur in a diploid individual. The interpretation of the locus/allelic relationships between multiple bands on an electrophoretogram where population data or only limited experimental data are available involves simple inspection of the phenotypes to create hypothetical genotypes which can then be tested numerically. For example, if three bands or fewer occur, designated a, b and c, and an array of phenotypes abc; ac; bc exists, it is simple to postulate that a and b are alleles at one locus whilst c is produced by another locus. The corresponding genotypes of the three phenotypic assays would be a:b c:c; a:a c:c; b:b c:c and this model could be tested against population distributions or segregation ratios from matings. If a more complex system of bands is observed, the joint exclusion of any two bands must indicate that they are not allelic unless a third allele exists. By inference, simpler models can gradually be built on the assumption that alleles are co-dominant and fully expressed. Null alleles can cause interpretive difficulty but not such as to preclude hypothetical genotypic models.

Finally, it is necessary that allelic frequencies are compatible with the observed frequencies of phenotypes and their postulated genotypes. This is the application of the famous Hardy–Weinberg law which was established independently by the two named geneticists shortly after the rediscovery of Mendel's work at the turn of the century. Indeed, had Mendel not worked with a self-fertilizing plant he may well have made the discovery himself, in which case we might now be calling it Mendel's third law!

The Hardy–Weinberg law states that in a panmictic population (defined below), there is a genotypic equilibrium determined solely by the frequencies of alleles and expressed as the expansion of the binomial $(p+q)^2$ where p and q are frequencies of two alleles A and B at a locus and $p+q = 1$. With three alleles, the binomial would be $(p+q+r)^2$ and so on.

For two alleles, the genotypic equilibrium is:

$$(p+q)^2 = p^2 + 2pq + q^2 \qquad (6.1)$$

where p^2 is the frequency of genotype AA, q^2 that of BB and 2pq that of the heterozygote AB. The simplest derivation of the law is that alleles segregate independently in the formation of gametes, random union occurs between spermatozoa and ova during fertilization and parents mate at random in the population. The first two conditions equal Mendel's first law; the third is the condition of panmixia, i.e. a population is **panmictic** if individuals can mate at random within it. Taken together, these three conditions imply that the population is a mixture of alleles which unite at random. That is, they imply that if allele A has a frequency of p in the population, the chance of two A alleles coming together during zygote formation is p × p or p^2, for two B alleles q × q, i.e. q^2, and for the heterozygote it is p × q and q × p, i.e. 2pq.

There is an important corollary of the Hardy–Weinberg law which is that whatever the mix of genotypes in generation 1, the equilibrium is established by one round of sexual reproduction and therefore is expressed in generation 2. So if two different populations of fish are mixed, the evidence for mixing disappears in one generation. The obverse of this is that if Hardy–Weinberg ratios are not observed, the existence of panmixia is in doubt and the population might be subdivided. An alternative explanation might be that certain genotypes are fitter than others and are favoured by natural selection. Such a process would normally lead to the near elimination of the alleles which confer the disadvantage but in some cases, e.g. superiority of the heterozygote, a balanced polymorphism is set up which maintains high frequencies for both alleles. This is an important realm of theoretical population genetics but its practical evaluation in fish is not well developed. The classical example in medical science is sickle cell anaemia, where the heterozygote is at an advantage by being partially resistant to malaria.

Finally, a further genetic perturbation may arise in the consideration of two or more loci. Where these segregate independently the full array of genotypes will reflect the random combination of those for each locus. Where different loci are linked the array will not be a random combination but biased – this is **linkage disequilibrium** and will be mentioned briefly in Chapter 7.

6.4 INFERENCES FROM POPULATION GENETIC DATA

The data on phenotype frequencies are processed arithmetically to generate allele frequency data, the total number of alleles scored being equal to twice the number of fish examined since each fish has two sets of genes. Thus in a sample of 20 fish, if six individuals are homozyous for allele A, four for the other allele B and ten are heterozygotes AB, the allele frequencies are simply $p=[(6 \times 2) + 10] \div 40$ for A and $q=[(4 \times 2) + 10] \div 40$ for B namely $p=0.55$ and $q=0.45$, respectively. A check on compliance with the Hardy–Weinberg equilibrium is then achieved simply by computing the terms p^2 $2pq$ and q^2, and multiplying them by 20, the sample size, to generate the theoretically expected numbers of individuals of genotypes AA AB and BB, respectively, which can then be assessed for goodness of fit against the observed numbers by appropriate statistical tests such as χ^2. In the hypothetical example the numbers would be 6.05, 9.90 and 4.05 – a remarkably good fit.

In addition to the allele and genotype frequencies three other genetic parameters are often derived from the observations, namely the frequency of polymorphic as opposed to monomorphic loci, the degree of heterozygosity over the loci studied and, where more than one population is analysed, the genetic identity of each population in relation to the others, i.e. the so called genetic distances.

Extent of polymorphisms

Not all soluble proteins show evidence of variation between individuals. Where such variation does occur, and is obvious even in small samples, the locus or loci responsible for the protein are described as **polymorphic** but clearly, some extremely low frequencies of alternate alleles will be difficult to evaluate and may, in any case, represent disadvantageous mutations, not genuine polymorphisms. Two levels of allele frequency have been invoked to define polymorphism. The least critical is that a locus is polymorphic if the frequency of the commonest allele is 0.95 or less – more critical is the level 0.99 or less. Clearly, in a sample of 50 individuals the chance of observing an alternate allele at the 1% level is not satisfactory for classification purposes, but this frequency and lower is commonly observed in relation to the incidence of recessive gene mutations of a disadvantageous nature. The decision to define a locus as monomorphic if its common allele is at a frequency greater than 0.99 is thus of doubtful validity. Unfortunately, there is little information on the fitness of homozygotes for rare alleles. By definition they can only be found in heterozygotes, and there is a need here for some conventional Mendelian crosses to be examined.

A very large number of soluble proteins has been assessed in fish. It is obvious that considerable differences exist between species, with some polymorphic for certain proteins which are monomorphic in others and vice versa. Some confusion does exist with respect to isozyme name and number and sample size may have an implication for the classification of polymorphism, but there is usually little difficulty in establishing an identification scheme for species based on protein electrophoresis.

A few proteins seem to be invariant in all of the species examined, but by and large the great majority are polymorphic in one species or another. The extent to which individual loci have alternate alleles seems to depend on species. The plaice, *Pleuronectes platessa*, for example, shows polymorphism at over 50% of loci examined, whereas loci in salmon are represented by alternate alleles in less than 10% of cases, and in bass, *Morone saxatilis*, and the paddlefish, *Polyodon spathula*, there appears to be very little isozyme polymorphism. There is no obvious reason for this variation between species and it does not appear to reflect a highly variable environment. In the case of the paddlefish, Carlson *et al.*, (1982) discuss several explanations of this low genetic variance without reaching a conclusion. Nevertheless the fact that some species are polymorphic for proteins and that others are not is a clear implication of natural selection at work. We shall discuss later the popular hypothesis that polymorphisms such as these revealed by electrophoresis are neutral in character.

Degree of heterozygosity

The frequency with which individual gene loci show variation in the form of multiple alleles is an indication of the genetic plasticity of a species. Another

Table 6.2 Electrophoretically detected allozyme variation and mean population heterozygosities for selected fish species for which large, unbiased populations were sampled

Species	Location (No. polymorphic)	No. of Loci	H	Reference
Scomber scomber	N. Atlantic	39(7)	0.049	Smith and Jamieson, 1980
Lates calcarifer	N. Australia	47(10)	0.032	Shaklee and Salini, 1985
Hippoglossus hippoglossus	N. Atlantic	43(4)	0.024	Fevolden and Haug, 1988
Micropterus salmoides	N. America	28(8)	0.040	Philipp et al., 1985
Pleuronectes platessa	Irish Sea	46(20)	0.102	Ward and Beardmore, 1977
Salmo trutta	British Isles	60(14)	0.038	Ferguson and Fleming, 1983
Salmo salar	Hatchery (landlocked)	38(3)	0.014	
	Baltic	38(3)	0.027	
	Baltic Hatchery	38(3)	0.020	
	East Atlantic	38(4)	0.026	Ståhl,1987
	East Atlantic Hatchery	38(3)	0.029	
	West Atlantic	38(5)	0.024	
	West Atlantic Hatchery	38(5)	0.015	
Cyprinodon (two species)	Texas	24.28(6)	0.048	Edds and Echelle, 1989
	Hatchery	24.28(6)	0.046	
Ptychocheilus lucius	Colorado	44(9)	0.043	Ammerman and Morizot,1989
	Hatchery	44(9)	0.047	
Polyodon spathula	Mississippi River system	35(2)	0.013	Carlson et al., 1982

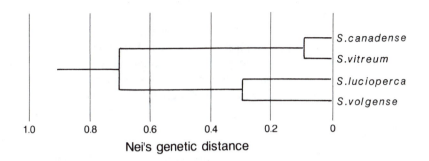

Fig. 6.3 A UPGMA cluster analysis of species within the genus *Stizostedion* (data from Billington *et al.*, 1991).

form of genetic variability in a population is the degree of heterozygosity (H), i.e. the average heterozygosity across all loci studied. Some examples of the two parameters are listed in Table 6.2, both have significance for population fitness but across different time zones. A lack of variation at the allele level is likely to be of long term significance in that evolutionary pathways are dependent upon the existence of such alternative gene forms. Lack of heterozygosity within a population reflects the extent to which alternate alleles are at high levels and this may be of greater importance to the well being of the population in the shorter term. We shall discuss the consequences of inbreeding in Section 6.6 but its main effect, reduction of heterozygote frequency, is debilitating. Whether lack of heterozygosity for a particular type of gene, such as those coding for allozymes, is indicative of heterozygosity over the whole genome is, of course, open to much doubt. Where H is used in relation to deliberate breeding plans, e.g. in close inbreeding, its interpretation is on much sounder ground since allozyme frequencies can be used as general markers.

Genetic relatedness

Whenever allele data are collected for individual populations or species or any other subdivision it is possible to ask the question to what extent do the groups differ from each other? This is of fundamental importance to the understanding of natural populations and taxonomy, and is also of use in defining captive populations such as domesticated strains or varieties.

Where only two subgroups are involved, a simple statistical test can establish whether or not the genotypic frequencies indicate that they are different – such tests do not prove identity, only difference. Where several populations are assessed it is also possible to do pairwise comparisons, but a much more powerful set of analytical tools is available to assess quantitatively the genetic distance between groups. The simplest of these measures is Nei's genetic distance (D) which can be presented as a dendrogram derived by the UPGMA method (unweighted pair group clustering method, Nei, 1975) in which the data are scanned initially for the smallest genetic distances and sequentially thereafter for progressively greater ones. The computation of large bodies of such data is now simplified by the availability of computer packages such as BIOSYS (Swofford and Selander, 1981), which also include provision for the other genetic parameters which may be derived from population data from isozyme studies, or indeed from any of the molecular methods to be outlined next. The output of such computer analyses on genetic distance for a species group and for a range of rainbow trout strains is presented in Figs 6.3 and 6.4.

Isozyme study has generated an enormous body of data on population genetics in the wild and in connection with captive stocks. Some of the implications of these studies for natural fisheries and fish farming will be presented in Section 6.7.

6.5 MOLECULAR STUDIES

Part of the tradition of the interpretation of isozyme data is the assumption of the existence of a 'biological clock'. This assumption is that the genetic material is slowly evolving by the progressive replacement of DNA coding sequences leading to changes in amino acid configurations in proteins. This results in independent evolution in reproductively separate groups and consequent divergence. For isozymes, this clock concept has been challenged on the grounds that the time intervals cannot be standard since selective forces can accelerate or retard progress. More reasonable claims for biological dating can be made for analyses of the structure of DNA which is not involved with the coding of proteins. This includes mitochondrial DNA and repeat-sequence

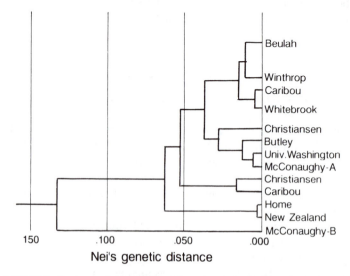

Fig. 6.4 A UPGMA cluster analysis of UK rainbow trout strains showing minor genetic variation (data from Thompson, 1985). Compare scale with that in Fig. 6.3.

DNA; in addition, non-transcribing RNA and histone protein essential to chromosome structural stability may also figure in future work of this sort.

Mitochondrial DNA

The use of mitochondrial DNA (mtDNA) for population genetic studies was reviewed by Ferris and Berg (1987). Mitochondria are intracellular organelles found in the cytoplasm and they contain small amounts of DNA. This is the only example of DNA existing naturally outside of the nucleus in eukaryotes. The function of the mitochondria is to regulate basic cell functions, but their principal interest in genetics lies in the structure of the DNA molecule within the mitochondrion. It is a circular molecule of double-stranded DNA comprising about 1800 nucleotide base pairs. Its structure is explored by the use of enzymes which cut the DNA at specific points to produce DNA fragments of diagnostic length.

The enzymes are described as restriction endonucleases; they derive from bacteria where their normal function is involved in the defence of the integrity of the bacterial DNA. They are named according to origin but classified according to the nucleotide sequences they recognize. There are thus 4-base, 5-base and 6-base cutters which recognize diverse but specific sequences of the adenine-guanine, cytosine-thymine base-pairs of the DNA molecule. Three examples are illustrated in Fig. 6.5.

After mitochondrial DNA is cut (digested) with restriction enzymes, the fragments can be detected by their mobility in electrophoresis. There is a rough correspondence between length of the fragment and mobility, but standards are available which can be 'run' with the digested mtDNA to corroborate

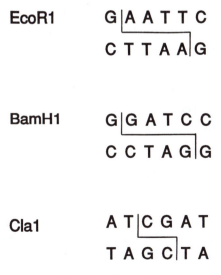

Fig. 6.5 Cutting sequences for selected restriction enzymes. Nucleotide bases: A, adenine; C, cytosine; G, guanine; T, thymine.

fragment size. A further control is that the sum of fragments must add up to the total length of the mtDNA molecule. Individuals show polymorphism for the array of fragments produced by restriction enzymes of which over 30 different sorts are now available. These are extremely expensive chemicals, however, and the need for such sophisticated methods requires careful justification.

These polymorphisms within populations can be assessed in quantitative terms such as genetic distance using exactly the same methods applied to isozyme variation. There is one important genetic distinction, however, between mtDNA restriction fragment length polymorphisms (RFLPs) and isozyme polymorphisms, in that the former are maternally inherited. They do not appear to be transferred with the male chromosome set at fertilization and therefore seem to have an existence which is independent of nuclear DNA. There are still some slight doubts about the universality of this model, but the practical consequences are that segregation and crossing over do not occur in mtDNA. Nevertheless, some evolution can arise by base substitution and the presence in an individual of more than one type of mtDNA molecule does happen (heteroplasmy).

In fish studies, extensive mtDNA variation has been observed in Atlantic salmon, *Salmo salar*, by King *et al.* (1992) and appears to indicate distant divergence between river systems as well as complex variation within individual rivers. Stock identity has also been shown for capelin, *Mallotus villosus*, by Dodson *et al.* (1991) across the North Atlantic Ocean and for widely spread populations of the anadromous ayu, *Plecoglossus altivelis*, in Japan (Pastene and Numachi, 1991). An important conclusion from the latter study was that huge efforts to enhance natural river stocks by the introduction of hatchery reared fish of land-locked origin over a period of 20 years had no genetic impact. In other words the introductions failed – such conclusions are not uncommon amongst the few cases where introductions have been monitored by careful genetic scrutiny.

Other DNA studies

The process known as DNA fingerprinting has been successfully applied in forensic work and has also been used for fish identification (Turner, 1991). The general procedures entail the digestion of extracted nuclear DNA with a mixture of restriction enzymes followed by electrophoresis which yields a very complex pattern of hundreds of bands which can be resolved into even greater detail by advanced electrophoretic technique. In general, however, this wealth of information reflects the uniqueness of individuals and is of little help in population work, though it could be relevant in the study of highly inbred or clonal material. An extension of it, to identify single components of the pattern by gene probe, which picks out specific genes by radioactive or other labelling, does bring order to the apparent chaos but the method then approaches the status and utility of other gene-orientated techniques such as the well established study of isozymes.

The DNA that separates out in the fingerprinting tests is probably from chromosomal intercalary heterochromatin (Chapter 7) and represents molecules comprising repeated specific sequences of nucleotide pairs. Another sort of DNA in rather larger quantities remains distinguishable on electrophoretoograms; this is described as satellite DNA and seems to be the material constituting the centromeric and telomeric heterochromatic regions of chromosomes. It is believed to be of structural importance. It comprises highly replicated repeat sequences which can be analysed by sequencing, i.e. by determining the exact structure of the nucleotide sequences. Such work in fish is of chief use in the wider taxonomic sense to identify distant relationships. Its use in short term population genetics and breeding is limited, as is the possible study of other macromolecules such as ribosomal RNA and the histone proteins which are specific to chromosomes.

6.6 INBREEDING

This topic is addressed in Chapter 9 but its close affinity to population studies requires it also to be dealt with here.

Inbreeding, the mating of close relatives, is a recognized tool in breeding but is also a hazard. As a tool, it has been widely used in the domestication of animals and plants in order to generate specific breeds which conform to certain criteria – often of visual impact rather than of economic significance. The gains are that inbred organisms are genetically more uniform than outbreds and that they have been 'purified' of harmful genes. The hazard arises because inbreeding leads to a generalized loss of fitness, so called inbreeding depression, which can be manifested in terms of survival, growth, fecundity and sundry other traits. The modern solution to these conflicting characteristics of the process of inbreeding is to use F_1 hybrids between two different inbred lines, thus eliminating inbreeding depression whilst retaining genetic uniformity. This topic in fish breeding will be covered in Chapter 9. The present section deals with the measurement of inbreeding effects in fish and its adverse effects.

Much theoretical work has been done on inbreeding and full texts are to be found in any book on population genetics. A simple account is sufficient for present purposes. Figure 6.6 shows the simplest and most intense form of inbreeding in gonochoristic species, brother–sister mating, with the alleles at one locus identified. In the grandparental generation the alleles at the locus are designated **a b c** and **d** for identification purposes, not to indicate absence or presence of allelic identity. The parents will contain **a** or **b** and **c** or **d** in the normal Mendelian way but the offspring will be able to inherit two of any of these alleles. The two alleles at the locus in question will then be *identical by descent*. The probability of this occurring is clearly dependent on the parental cross, thus ac × ad, as in Fig. 6.6, leads to one-quarter of offspring being homozygous by descent whereas ac × bd gives none and ac × ac gives one- half. Overall, the proportion of pairs of alleles identical by descent will be

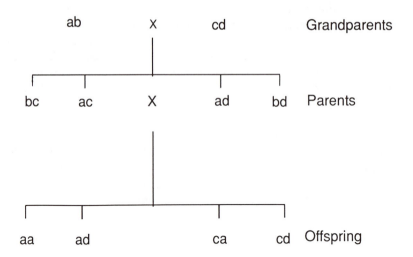

Fig. 6.6 Schematic representation of the genetic consequences of inbreeding. See text for interpretation.

one-fourth. If an allele frequency in the grandparental generation is p, the frequency of its homozygote by descent in grandoffspring will be p/4. Under normal conditions of panmixia, the frequency of homozygotes would be p^2, but since we have one quarter already identical by descent this reduces to $0.75\,p^2$. The difference between inbreeding and panmixia for any homozygote for one generation is:

$$p/4 + 3/4\,p^2 - p^2 = p/4\,(1{-}p) = pq/4$$

The effect of inbreeding is to change genotype frequencies in a way which is independent of allele frequencies and does not change population allele frequencies themselves.

These arguments assume that populations are large and that inbreeding may lead to a range of inbred lines amongst which all of the initial allelic variation is still to be found and from which, by crossbreeding, the initial population structure could theoretically be recreated. There is, however, a corollary to this in that if only small numbers of inbred lines are maintained – the usual practice – the fixation of one specific allele at a frequency of 100% in a line by the chance loss of other alleles constitutes loss of genetic variation. This is not a consequence of inbreeding itself but of so called *genetic drift*, the stochastic loss of specific alleles when the number of parents chosen for the next generation is small.

A further general point is that hybrid vigour or heterosis as seen in F_1 hybrids is an almost invariable consequence following the crossing of two unrelated inbred lines and since, by genetic drift, neither can have more than a fraction of the allelic variation of the original population, the full fitness observed is due to generalized heterozygosity, not to any specific allelic array within the F_1 individual. There have been many reports of heterozygote advantage for specific isozyme loci (Kirpichnikov, 1991). These frequently take the form of flexibility of function at variable temperatures, and whether or not this specific effect is distinct from or part of heterosis is not yet established. What does seem clear is that such specific examples of single locus effects are unlikely to be numerous in one population. If a genotype has a specific advantage it follows that the alternatives are disadvantaged. For such disadvantages to arise from several loci in a population would imply a heavy level of genetic disability. This is the old paradox of genetic load, first enunciated by H.J. Muller, which is still unresolved except in the generalization that heterosis is nonspecific and not due to heterozygosity at specific loci.

Inbred lines for some aquarium fish have been reported but data on performance are limited. In the convict cichlid, *Cichlasoma nigrofasciatum*, Winemiller and Taylor (1982) observed no major deleterious inbreeding effects until the F_4 or F_5 in a line started with presumed sibs from a local pet shop. A similar result was obtained with Guppies starting with specimens from a feral population (Purdom and Scholes, unpublished). Seven lines were started. No obvious loss of viability was found at F_6, when inbred fish were compared with those from the base population retained in mass culture, but severe inbreeding depression was obvious from F_9 onwards and all but two of the original lines had ceased to exist. The outcome of a schedule of inbreeding is very dependent on the nature of the starting material, particularly any prior history of inbreeding. The number of lines at the outset is also crucial – the more the better – and intervening generations of mass mating could give greater stability to the survivors once inbreeding losses start to take a toll of individual lines. There is much scope for varying the procedures of inbreeding schedules if a stable inbred line is being sought. Monitoring of the whole process by electrophoresis or by external appearance, as is possible in the Guppy and the platy, is highly desirable.

In commercial food fish, the existence of heterosis in inter-strain crosses of common carp (Moav *et al.*, 1975) is indirect evidence of inbreeding depression in the strains themselves. In salmonids, Kincaid (1983) reports significant levels of inbreeding depression across a range of characteristics. The comparison of commercial strains of rainbow trout with their hybrids (Hörstgen-Schwark *et al.*, 1986) did not reveal heterosis for performance characteristics, and similar results have been obtained by me with a wide range of rainbow trout strains that are used in the UK. Despite hearsay evidence of minor levels of inbreeding and the existence of very significant differences amongst the strains themselves, the hybrids showed intermediate levels of performance. Inbreeding, as with other aspects of population genetics in fish, is well provided with theoretical background but not with evidence from practical application. The adverse consequences of inbreeding are probably overstated, not least because natural selection is a powerful force against the maintenance of homozygosity. The complexities of breeding schemes to avoid it whilst maintaining a selection programme are discussed by Hershberger *et al.* (1990a). Simpler systems with maximum broodstock numbers are more practicable, however, and unlikely to lead to inbreeding problems which cannot be solved by an occasional introduction of unrelated stock.

6.7 APPLICATION OF ELECTROPHORESIS

The central question asked in virtually all studies of electrophoretic variation in fish is to what extent one population or sample is different from another. In a slightly different form, the question may be whether the 'population' is a single entity or divided into distinct subunits.

The reasons for asking this question are numerous:

1. academic: e.g. do the putative populations reflect developments of evolutionary significance or of shorter term ecological importance;
2. for fisheries management: e.g. are individual stocks discernible for which different management practices could be envisaged, are lesser stocks subject to over-exploitation from fishing effort based on a bigger stock;
3. for aquaculture: e.g. do different natural stocks exist from which domesticated strains can be derived, do existing farmed stocks differ genetically amongst themselves;
4. for fish health: is there evidence of adverse change in a farm stock from undue inbreeding or unplanned genetic mixing;
5. environmental: are natural stocks in danger because of loss of genetic variance or by disruption of natural co-adapted gene pools by introgression of intra- or interspecific hybridization?

Examination of fish by electrophoresis of soluble proteins or DNA fragments or by immunological methods has been carried out for population study on a massive scale world-wide. The number of reported applications, however, is very small. One of the reasons for this disparity is that information on the genetic structure of fish populations is unlikely to be of use unless specifications are made, in advance, of the problem and the nature of the information needed to solve it. There appeared to be a generalized expectation in the 1960s that an awareness of the genetics of fish populations would provide the basis for the solution of fishery problems as they came along. This view still prevails in some quarters and has in part been responsible for the pursuit of ever more sophisticated methods of population analysis. No single general philosophy on population genetic structure has emerged, however, from this huge programme of work, but a wide range of fish population characteristics has been revealed and these should be taken into account more fully and not ignored in the pursuit of yet more novel ways of looking at the problems.

Divergence of populations

It is almost always the case that if two populations of a species of fish are separated by clearly defined barriers to introgressive hybridization, a population study by biochemical or molecular biological means will show evidence of divergence. Atlantic salmon is an excellent example. The early work of Payne and Child (1971) demonstrated population divergences in Great Britain which could be related to the retreat of glaciers from the Würm ice age. Later work, summarized by Ståhl (1987), has shown the wider divergence between three groups representing North American, European and Baltic races of salmon and progressively more finely based differentiation into rivers and even tributaries (Davidson *et al.*, 1989). The generating force behind this remarkable divergence is believed to be the precise homing of maturing fish to their natal home. How this accommodates the known propensity of these fish to colonize new territory in the space of a few thousand years is not so easy to perceive. Similar but even greater divergences have been reported many times for populations of brown trout in Europe and rainbow trout and brook trout in the United States.

This finely tuned differentiation of populations presents a problem in fishery management or in conservation terms, simply because identifiable units seem to arise wherever one looks and arbitrary lines have to be drawn somewhere. Less troublesome are the situations with more broadly based population identity such as within the striped bass, *Morone saxatilis*, of the Atlantic coast of America and the Gulf of Mexico. Wirgin and Maceda (1991) describe nuclear DNA probes which differentiate subpopulations of these fish in a way that isozyme studies have so far failed to do. This type of population discrimination of non-interbreeding populations can be used for fishery management purposes. The situation is likely to be found more often in coastal fisheries of anadromous species than in the open sea or with purely marine species.

Undifferentiated populations

These are commonly observed where geographic barriers appear not to exist; high seas fisheries are obvious examples. The tuna fish species provide classic examples, from the earlier blood-grouping studies up to modern day DNA technology (Graves *et al.*, 1984). Similarly bland populations are found across the natural distributions of hake, *Merluccius merluccius*, marine flatfish such as plaice, *Pleuronectes platessa*, in the North Sea (DeLigny, 1969) and halibut, *Hippoglossus hippoglossus*, (Fevolden and Haug, 1988) across the North Atlantic – even though return migration to discrete spawning sites is well established for some of these species.

Each of these examples includes putative populations but the consensus is that pelagic drift of larvae permits active gene flow between subgroups and prevents divergence. This seems a not unreasonable explanation but it has an important corollary, which is that the overall genetic structure of the population represents a compromise. This is not compatible with the view that coadapted gene pools are vital to the success of populations. The lack of genetic identity of these populations is not helpful to the management of the fisheries. The level of intermixing needed to prevent genetic divergence is of the order of the reciprocal of the population size, and this is far too low to have short term fishery implications.

Clines

Clines are especially interesting in population genetics. They comprise a progressive change in allele frequency along a gradient which can usually be defined in environmental terms such as temperature or salinity. The implication is that the polymorphism is adaptive and that the genotypic frequencies reflect the differential fitnesses of individuals across the gradient. Classical examples are found in Catastomids (Koehn, 1970) and largemouth bass (Philipp *et al.*, 1985). An experimental analysis has been made in *Fundulus heteroclitus* which supports a model of a natural cline for lactate dehydrogenase alleles along a natural temperature gradient (Powers *et al.*, 1991). Much wider ranging clines have been reported for species such as the cod, *Gadus morhua*, but the ecological background is not so well established as in the smaller fish species.

Divergence of sympatric species

This is an unusual type of population differentiation. The best studied example is that of three sympatric populations of brown trout, *Salmo trutta*, in Lough Melvin in Ireland (Ferguson and Mason, 1981). These populations were identified as distinct genetic entities on the basis of allele frequencies at five polymorphic enzyme loci. They were already known to fishermen as distinctive types of brown trout by their external appearance. A brightly coloured form with obvious red spots was known as the gillaroo, a darker leaner fish was called sonaghen and the third, a larger dull coloured fish with the appearance of a top predator, was called ferox. A fourth type of trout recorded in the Lough, described as brown trout, seems to be regarded by Ferguson and Mason as either a relic from previous casual stocking from elsewhere, or simply as a catch-all of fish not classified as belonging to one or other of the other three.

The three distinctive types were genetically distinct and their genetic isolation was explained on the basis of their different life styles – mollusc eating, planktivorous and piscivorous, respectively – and their separate spawning grounds. There is therefore almost a case for regarding them as allopatric. However, they do occupy the same body of water, and Ferguson and Mason report concern that further stocking of alien brown trout might destroy the basis for their separate existence. The same effect could follow an environmental change either to the Lough, itself e.g. eutrophication and a change in prey availability, or to the spawning beds.

6.8 SUMMARY

The evidence from population genetics is therefore equivocal. That genetic adaptation occurs is indisputable, but that it is uniquely essential for population stability is obviously doubtful.

Stock enhancement tactics

The existence of genetically distinct stocks raises the question of what approach to take for the management of stock enhancement programmes. With increasing demands on diminishing fish resources, the use of enhancement practices seems bound to increase. The obvious solution is to rear stock local to the region where release is to take place, but this is not always possible nor is it necessarily the right approach. The fact that the fish for release are hatchery reared can be expected to introduce new determinands on what constitutes an appropriate stock. There is a reasonable case for constructing as diverse a stock as possible by outcrossing known geographically isolated populations. At least with an enhanced level of genetic variance, the new population will have a greater potential to adapt to the new circumstances. The conservation aspect of such a choice might, however, be suspect.

A common conservation issue is the extent to which natural populations can be affected by deliberate or accidental introductions of hatchery reared fish of the same species but possibly of different genetic backgrounds. This problem is discussed in Chapter 9 in relation to introgressive hybridization, but a comment is also relevant here in connection with population genetics.

Conservation fears

The conservation issue for Atlantic salmon is frequently raised but not resolved. Salmon and trout intraspecific enhancement exercises have been attempted casually or purposefully on very many occasions but rarely with any monitoring back-up. The most common result where studies have been undertaken is that the introductions were ineffective in changing the genetic structure of natural populations and possibly ineffective as introductions per se. It is probably more rational to use a genetic monitoring approach to assess the success of the stocking policy than to assess any impact on natural population genetics.

A slightly different problem arises with the striped bass situation, where large but partly overlapping discrete stocks are found. To what extent would introgression destroy the mechanisms which perpetuate the differences?

The most cogently argued case over a genetic consequence of introductions is the work by Chilcote *et al.*, (1986) on the success of hatchery reared steelhead trout, *Oncorhyncus mykiss*, and further debated by Campton *et al.*, (1991) and Chilcote *et al.*, (1991). The original work entailed the release of hatchery steelhead homozygous for a rare isozyme allele and the recording of returning adults (these are anadromous trout) and their offspring by scoring fish for the allele in question. One contentious result was the demonstration of differential fitness of different isozyme genotypes but this is not pertinent here. The issue of concern for introduction policy was that the reproductive capability of the hatchery returning adults was very much less than that of the natural returning adults. Despite the depth of the treatment accorded to this experiment it was still not possible to distinguish between effects due to the life history of the hatchery fish (i.e. environmental) and those due to their genetic composition, as acknowledged by both sets of authors. What seems clear is that once interbreeding begins, the genetic composition of F_2 and later generations is no longer defined by the isozyme locus. Further light would have been thrown on the subject had the analysis also included mitochondrial DNA studies to identify parents.

A similar situation exists for the Australian barramundi, *Lates calcarifer*. It is a coastal fishery of an anadromous species and individual, genetically distinct, geographic populations have been identified (Shaklee and Salini, 1985) by conventional protein electrophoresis. Finally, at an even greater level of proximity, there is the question of what might happen to a group of sympatric populations of the sort found in Lough Melvin. It might be argued that the forces which keep all of these situations in a stable form are strong enough to withstand introgression. This seems more logical than the fear that an introduction would disrupt the natural balance, but the risks attendant on a deliberate experiment *in situ* to find out seem too great to warrant the attempt.

Hard evidence in this area is difficult to obtain but Skaala *et al.* (1990) review the data on interaction between farmed and natural species for a range of fish species and conclude that "... evidence for genetic interactions ... have been found in a few cases". More evidence is needed in this contentious area and negative results should not be disregarded.

Classical population genetic theory deals with the fitness of individuals in populations. Conservation is concerned with the fitness of populations. There is much confusion over the distinction between these two ideas about fitness and most of the concern expressed over the potential genetic harm which could arise from population admixtures stems from an incorrect application of classical population genetic theory. Whilst a population maintains an effective mode of sexual reproduction and a high level of potential recruitment, the adverse genetic effect of newly introduced variation will be manifested as genetic damage to individuals. Their removal from the population is the generative force of evolution and it leads not to population genetic damage but to the reverse, population recovery or advance! However, until this controversy is resolved and criteria are established for the proper evaluation of the population genetic models, the popular concern over introductions and escapes is likely to remain a limitation on fish farming, ocean ranching and the concept of stock enhancement.

Fishing impacts

To what extent commercial fishing represents a potential selective force depends on methods. Law (1991) has examined the potential effect of size selection in fisheries and concluded that it could be responsible for observed reductions in age at sexual maturity for NE Arctic cod. Similar considerations for Pacific salmon were also discussed but the requirement for heritability values for growth rate to be as high as 0.3 render such arguments invalid.

The problem with natural fisheries is that their dynamics is not sufficiently stable, nor sufficiently well understood, for predictions to be made and tested on population genetic models. Some characteristics of fishing, such as the seasonality of exploitation of salmon, seem more likely than others to influence future generations but the analysis of them is very long term and speculative. Genetic studies of the recovery of overexploited populations would also seem to be a feasible aim – to answer the question, for example, of whether recovery or recolonization has occurred. Finally, some evidence of population structural changes due to fishing pressure has been obtained (Smith *et al.*, 1991), but the genetic implications of such events cannot yet be assessed critically.

Chapter seven

Chromosomes of fish

7.1 INTRODUCTION

Although genes display the mechanisms of inheritance by the ways in which heritable characteristics are expressed in successive generations, it is the chromosomes which determine these mechanisms. This was the basic simple logic which, in 1902, led W.S. Sutton to the hypothesis that the chromosomes provided, in fact, the physical basis of heredity. This brilliant insight was immediately rewarded and the chromosomal basis of the laws of inheritance came into being and dominated genetics from its early beginnings and even up to the present, molecular, age.

What are chromosomes? They are named from the fact that they can be stained preferentially with certain dyes and so show up in suitable cellular material prepared for microscopical examination. Their detailed structure is still not fully understood but they appear thread-like at certain times and are known to be composed of linear complexes of deoxyribonucleic acid (DNA), the genetic material proper, and histone proteins which are believed to have a supporting or structural role.

All living organisms have chromosomes, but those of the bacteria and viruses (prokaryotes) are somewhat different from those of plants, unicellular and multicellular animals (eukaryotes) and will not be considered further. The cells of eukaryotes characteristically have a well-defined cell nucleus which carries the chromosome complement of each cell. The number of chromosomes per cell is characteristic for a species although in fish we shall see later that some variation can be observed between members of the same species and even between cells in one fish. The chromosome number in the cells of the body is normally made up to two sets, one of maternal and one of paternal origin, and the double state is described by the term diploid. Within the animal kingdom the diploid number of chromosomes varies from two in some nematodes to over 200 in certain species of fish. A single set of chromosomes as found in a mature egg or in a spermatozoan is described as a haploid set, and where more than two sets of chromosomes occur in cells, the condition is termed polyploid, or more specifically, triploid for three sets, tetraploid for four sets and so on.

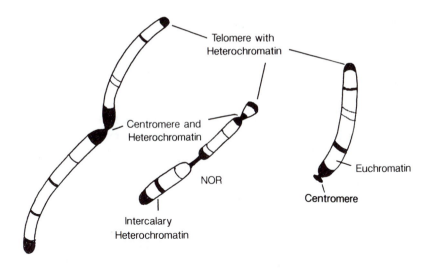

Telomere with
Heterochromatin

Centromere and
Heterochromatin

NOR

Euchromatin

Centromere

Intercalary
Heterochromatin

Metacentric   Acrocentric   Telocentric

with Nucleolar
Organizing Region

Fig. 7.1 The basic chromosome structures with location of main heterochromatin
bands and nucleolar organizing region.

7.2 METHODS IN CYTOLOGY

The basic structures of chromosomes as detected under the light microscope
are shown in Fig. 7.1. The material of chromosomes is called chromatin and
two sorts are recognized – euchromatin which stains but lightly and heterochromatin which stains darkly. This differential staining is the basis for most
of the methods now used for detecting details of the structures of chromosomes. What it reveals is the degree of condensation or spiralling of the DNA
thread. The classical double helix structure for DNA is its functional aspect,
giving scope for replication of itself and for the accurate transcription of DNA
as the first step in protein synthesis. But in observed chromosomes the
DNA thread is very highly condensed and the total length of a chromosomal
DNA molecule is many thousand times the length of the chromosome as
observed during cell division. It is the mechanics of the division cycles which
requires the condensation of chromosomes and makes them observable. The
different phases of the cycles of mitosis and meiosis are characterized by
different chromosomal appearances.

Euchromatin is believed to contain the genes in a linear array like beads
on a string, whilst heterochromatin is regarded as genetically inert and to
have a function in maintaining the structural integrity of the chromosome
and perhaps regulating gene expression. Heterochromatin is mostly made up

of highly repeated simple sequences of DNA. The genetic DNA itself comprises long chains of sugar–base complexes linked by phosphate bonds and existing in a two-strand form in which specific base pairs are cross-linked (Chapter 2). There are four bases, the cross-linked pairs being cytosine–guanine (CG) and adenine–thymine (AT). The base pairs provide the code for the manufacture of proteins, in which triple sequences of base pairs code for individual amino acids and act more or less as a template to generate first a ribonucleic acid (RNA) copy of the template which in turn lines up the amino acids to construct a protein. It is beyond the scope of this book to go into greater detail of genetic transcription processes, which can be seen in most texts on molecular biology or modern genetics.

Two other features of chromosomes in Fig. 7.1 are the ends, called telomeres, and a constriction called the centromere. The telomeres are special in the sense that they are stable entities, usually heterochromatic, which 'round-off' the chromosome thread. If chromosomes are broken, as for example by the passage through them of X-rays, the broken ends are 'sticky' and tend to re-join but not always in the original way. Telomeres are essential for maintaining the integrity of chromosome threads.

The centromere controls the movement of chromosomes during cell division. After the chromosome threads have replicated, the daughter chromosomes move apart under the control of the centromeres. Like the telomeres, centromeres appear to be located in heterochromatin. Of more obvious importance, however, is the position of the centromere along the length of the chromosome. When it is located near to the centre, the chromosome has two clearly discernible arms and is described as metacentric. Rod-shaped chromosomes have the centromere at or very near to one end and are called telocentric, whilst acrocentric chromosomes are those with one arm very much longer than the other. The prefix 'sub' is sometimes used to described intermediate cases. In addition to the major gross features of chromosomes, sundry other constrictions may be seen following simple staining, and a feature of one or more chromosome pairs in a nucleus may be a nucleolar organizer region (NOR). Such details became more apparent with modern sophisticated staining procedures (p. 117).

Fish chromosomes are not easy to study. They are usually very numerous and often small and lacking individual characteristics. Early studies were bedevilled by poor techniques which were barely adequate to do much more than demonstrate the presence of chromosomes. Thus over the period from 1920 to 1945 Svardson (1945) lists chromosome number estimates for brown trout, *Salmo trutta*, ranging from a diploid number of 24 up to 80. Modern cytogenetic techniques (Hartley and Horne, 1985), which will be covered fully later, confirm the true diploid number for brown trout as 80. Counts of chromosome arms are even more fraught with uncertainty, possibly because the contraction, or shrinkage, which accompanies fixation prior to staining may vary considerably with different techniques or even within a technique used in different laboratories. Nevertheless, the number of arms or 'nombre fondamental' (NF value) is extremely important in any consideration of chromosomes in evolutionary or breeding contexts since this defines the genetic content of a chromosome complement.

Before proceeding further with a discussion of the chromosomes of fish, it is necessary to return to the general behaviour of chromosomes in cell division, since these events are crucial not only to an understanding of cytological method and of the role of chromosomes but also to an appreciation of practical manipulations of chromosome sets (Chapter 12).

Mitosis

This cell division simply produces two identical 'daughter' cells from one original cell. The genetic material in the chromosomes is replicated and each daughter cell receives a set of exact copies.

Whilst actively transcribing RNA for protein synthesis, the chromosome threads are in a diffuse phase and difficult to observe under the microscope. The nucleus in the so-called resting state (interphase) seemingly lacks structure. In preparation for division, the chromosomes become constricted, possibly by multiple coiling, and observable as discrete entities. This is the stage of division called prophase and during it the DNA threads become replicated base for base along their lengths; prophase ends with the disappearance of the nuclear membrane and the start of the period called metaphase during which further condensation of the chromosomes occurs and they become orientated on the metaphase spindle (Chapter 2, Fig.2.2). The replication of the chromosome to produce two daughter chromatids is not accompanied, however, by division of the centromere, so the chromosomes from the end of prophase and into metaphase are doubled and appear V-shaped if telocentric and cross-shaped if metacentric or acrocentric. The metaphase spindle, a complex of microtubules, appears like two cones of fibres with bases opposed. The chromosomes come to lie in the region of the opposed bases (the metaphase plate), the centromere divides, and each new chromosome of a pair moves away from its partner and towards the appropriate apex of the cone. This occurs more or less at the same time for each of the original chromosomes so that the metaphase division evenly divides the replicated chromosome set back to the prereplicated condition, and the remaining stages of anaphase and telophase see the new chromosome sets aggregated at the poles of the spindle, enclosed within a new nuclear membrane, and put back to work.

At the time when they orientate at the metaphase plate, the chromosomes are in their most condensed phase and this is why this particular part of the cell cycle is preferred for most cytological study. Because they are so condensed, the chromosomes are also likely to be easily spread at this time, which facilitates counting and classifying. There are various ways of looking at the spindle, however, and polar views tend to be less informative than equatorial views. Whatever the view, the presence of v-shaped or x-shaped chromosomes indicates that the mitotic cycle is in metaphase and careful scrutiny of a prepared slide will reveal those 'spreads' which are scorable and those which are not.

Meiosis

This type of cell division was briefly described in Chapter 2. It is, in effect, two cycles of division with only one phase of chromosomal duplication such that it produces four haploid products from one diploid cell.

In the first part of the cycle, meiosis I, prophase is a long complicated stage in which the chromosomes first appear to contract, as in mitosis, but then immediately pair up, with maternal and paternal chromosomes coming together precisely along their lengths. DNA replication then occurs, making the paired structures four-stranded, but the centromeres do not divide. Metaphase follows, but still the centromeres remain undivided, so that anaphase and telophase embrace double structures resembling paired daughter chromatids as in mitotic metaphase. No further DNA replication occurs in meiosis II but the centromeres divide during metaphase II, and telophase II sees the halved number of chromosomes appropriately segregated to the opposite poles of the spindle. In meiosis in male animals, one diploid spermatocyte is transformed into four haploid spermatozoa whilst in the female, one set of chromosomes is denatured after meiosis I – the first polar body, and one after meiosis II – the second polar body. The final effect is to produce a haploid set of chromosomes in both egg and sperm.

Two further features complicate meiosis. One is simple, the other complex. The simple feature is that maternal and paternal chromosomes – or rather their centromeres – are randomly accumulated at the spindle poles in metaphase I and thus maternal and paternal complements get thoroughly mixed up. This is the physical basis for Mendel's second law of independent assortment (p. 32).

The second, more complex, feature, occurs during replication of the chromosomes in prophase I and involves exchange of material between a daughter chromatid of maternal origin and its partner of paternal origin. This is the process of crossing over or recombination and is believed to occur by imprecise linking of DNA sequences during the replicating phase in prophase I.

One consequence of crossing over is that the homologous chromosome pairs, which constitute four-stranded structures once replication has occurred, are linked together such that when the unreplicated centromeres move apart after metaphase I, the double structure takes the form of a cross or chiasma (Fig. 7.2). As anaphase I unfolds the chiasmata slip towards the ends of the chromosomes and eventually the chromosome arms become free of each other. Chiasmata are important indicators of maternal/paternal chromosome exchange and they may also have a function in stabilizing the paired chromosome structures at metaphase I and permitting regular segregation during anaphase and telophase. If this segregation breaks down for a chromosome pair, the consequence would be that both chromosomes or neither would end up in the new daughter nucleus. This is called non-disjunction and has important genetic implications, (a) as a generally deleterious accident and b) as a provider of 'extra' genetic material.

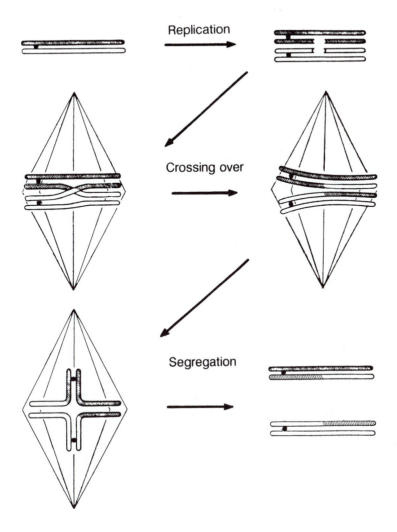

Replication

Crossing over

Segregation

Fig. 7.2 Crossing over and its relationship with chiasma formation.

The chromosomal patterns of meiotic divisions are much more difficult to assess than those of mitotic metaphase but because of the pairing and the chiasmata, they can be much more revealing than mitotic studies.

Tissue techniques

The basic requirement for a tissue for chromosome study is, of course, that it contains cells undergoing division. Much of the early work with fish used early embryos for mitotic study (Svardson, 1945) and male gonads for meiotic material (Nogusa, 1960). The greatest advance in fish cytogenetics, however, came after the development of techniques using blood tissue culture systems.

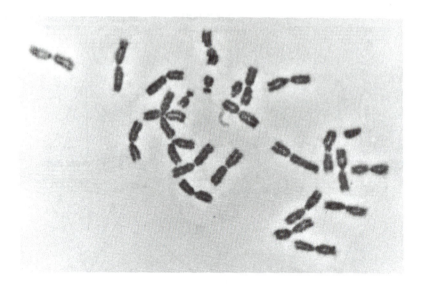

Fig. 7.3 Chromosomes of the cyprinodont *Ameca splendens*: 2n = 26. (Courtesy of Dr D.S. Woodhead).

These methods were first used for human chromosome studies; they depended on the remarkable action of certain plant materials, phytohaemagglutinins (PHA) as discussed in Chapter 6, in inducing cell division in circulating lymphocytes. The full technique (Evans, 1965) involved extracting a small volume of blood, using heparin to prevent clotting, inoculating it into a tissue culture medium containing PHA, and incubating this at appropriate temperature for a few days. After this, the cells were treated with colchicine, which blocks the mitotic cycle at metaphase, fixed in accordance with standard cytological technique (Darlington and La Cour, 1950), spread on a microscope slide by air drying, and stained (acetic orcein) and permanently mounted. The quality of the chromosome spreads achieved by such methods marked a major advance in animal cytogenetics (Fig. 7.3).

Analysis of chromosomes using blood leucocyte culture in fish was performed by many cytologists once the obvious advantages of these techniques for mammalian study became apparent in the mid 1960s. In general, the rates of mitotic divisions in peripheral blood of fish are sufficient without stimulation by PHA (Barker, 1972), but this remains a usual feature of technique. The use of other tissues such as kidney and gill epithelium also benefited from the post-mitotic procedural advances, particularly in relation to colchicine treatment (Kligerman and Bloom, 1977), air-drying to achieve good spreads (McPhail and Jones, 1966), and sundry other aspects of technique, particularly in salmonids (Chourrout and Happe, 1986) and in cyprinids (Gold *et al.*, 1990).

Staining techniques

Overall staining of chromosomes is achieved with dyes which have a general affinity for DNA, but finer detailed resolution of chromosome structure is revealed by a variety of chromosome banding techniques. These methods largely depend upon the distribution of heterochromatin as revealed by partial denaturing of the chromosomal DNA which differentially affects various components of euchromatin.

Fluorescence banding

Q-bands are achieved by staining with quinacrine compounds followed by observation via ultraviolet microscopy (Caspersson *et al.*, 1969).

Constitutive heterochromatin banding

C-bands are generated by the vigorous or alkaline hydrolysis of prepared chromosome spreads followed by Giemsa staining (Comings, 1978). Solid blocks of heterochromatin remain and take up stain whilst the rest of the DNA is removed or denatured by the hydrolysis. The effect is mostly to indicate the presence of large heterochromatic blocks at the telomeres and on either side of the centromere in acrocentric or metacentric chromosomes. Recent developments suggest that digestion with restriction enzymes (Chapter 13) may give additional resolution to C-banding in fish chromosomes (Hartley, 1991).

Giemsa banding

G-bands are produced by controlled alkaline hydrolysis of chromosomes prior to staining with Giemsa (Pardue and Gall, 1970) or by tris borate buffered acid hydrolysis (Sumner *et al.*, 1971). This treatment is less harsh than that for C-banding, and presumably leaves behind not just the solid blocks of heterochromatin but that in the more varied interstitial zones along the lengths of chromosomes.

Nucleolar organizers

NOR staining is achieved with silver salts followed by differentiation with sodium thiosulphate (Goodpasture and Bloom, 1975). NORs can also be recognized after fluorescence banding.

Late replicating DNA

Incorporation of tritiated thymidine followed by autoradiography can differentiate between early and late replicating DNA (Caspersson *et al.*, 1969) and has been used to demonstrate this property in some of the major blocks of heterochromatin already identified by fluorescence microscopy.

The deployment of the range of banding and staining techniques has permitted the clear identification of each of the 23 chromosome pairs in Man. Figure 7.4 shows the degree of differentiation now achievable in medical science. The chromosome spread is of the trisomic form of Down's syndrome

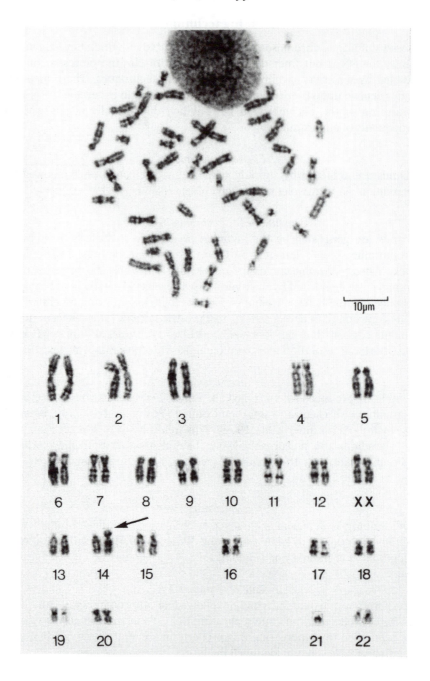

Fig. 7.4 Chromosomes of Man – a spread of the chromosome complement in Down's syndrome showing the extra chromosome (arrow). (Courtesy of Professor H.J. Evans).

10 μm

Fig. 7.5 Chromosomes of the Atlantic salmon. (Courtesy of Dr S.E. Hartley).

and the extra chromosome is clearly defined and identifiable as are all the others. Figure 7.5 shows a good example of current technology in fish chromosome preparation and despite the clear identification of the 58 chromosomal elements (*Salmo salar*), the detail is not yet up to that shown in human chromosome preparation. There seems no reason why one day the fish cytologists should not reach the same level of achievement now possible in medical cytology. The study of chromosomes in fish has not been subjected to the same comprehensive technical appraisal that has been applied to human studies. While some advances have been made, the detailed nature of chromosome variation in fish and its significance in breeding remains largely to be discovered.

7.3 CHROMOSOME NUMBERS AND GROSS MORPHOLOGY

A wide range of taxonomic groups of fish has been studied and much speculation made about the evolution of fish karyotypes. Table 7.1 lists chromosome numbers from a variety of fish species and also total DNA contents of the nuclei. The enormous taxonomic variation is obvious and some generalizations can be made. Ohno (1974) traces fish evolution from primitive states 500 million years ago, roughly equivalent to tunicates amongst modern fauna, to today's great variety of species and genomes. The amazingly

Table 7.1 Chromosome numbers, arm numbers and DNA contents of a range of fish species*

Species	2n	NF	relative DNA	Haploid size $g \times 10^{-12}$
Pleuronectes spp	48	48	0.20	0.7
Oreochromis niloticus	44	58	0.27	1.0
Esox lucius	50	50	0.39	1.4
Umbra limi	22	44	0.72	2.5
Dallia pectoralis	78	94	0.36	1.3
Carassius auratus	104	166	0.49	1.6
Cyprinus carpio	104	150	0.50	1.8
Barbus barbus	100	144	0.49	1.7
Barbus capensis	150	208	–	–
Barbus fasciatus	50	82	0.22	0.8
Channa punctata				
var.A	34	64	–	–
var.B	32	64	0.19	0.7
Channa gachua	78	104	0.28	1.0
Oryzias luzonensis	48	96	0.54	1.9
Oryzias latipes	48	68	0.54	1.9
Salmo salar	58	74	0.86	3.0
Oncorhynchus kisutch	60	108	0.90	3.1
Coregonus lavaretus	80	96	–	3.2
Thymallus thymallus	102	170	0.60	1.9
Mammalian	–	–	1.0	7.0

*Sources: Banerjee *et al.* (1988); Beamish *et al.* (1971); Dhar and Chatterjee (1984); Formacion and Uwa (1985); Hartley (1987); Hartley and Horne (1984a); Majumdar and McAndrew (1986); Ollerman and Skelton (1990); Ohno (1970, 1974).

well-endowed genomes of the lungfishes are regarded as secondarily derived and probably due to lead nowhere, i.e. they occupy an evolutionary cul-de-sac. On the other hand evolutionary progress can be discerned in taxonomic groups with small genomic or chromosomal contents. Progress is achieved either by evolution of the basic structure of chromosomes or by polyploidy, which can generate scope for further mutational change and evolution. Certainly, the higher groups of fishes such as the Perciformes have relatively small chromosome numbers and DNA contents, whereas the more primitive forms such as the Salmoniformes have large chromosome numbers and DNA values to match. The possibility that salmonids are recently arisen tetraploids is a recurring theme (Ohno, 1974).

A similarly wide-ranging review by Kirpichnikov (1981) of the evolutionary significance of karyotypes and genome size in fish reflected similar conclusions. Kirpichnikov assessed the material stepwise, going from the cyclostomes, poorly documented but having two quite disparate groups in the hagfishes and lampreys, respectively, to teleosts via elasmobranchs, dipnoi and sturgeons. All of these latter fishes are regarded as of primitive origin in

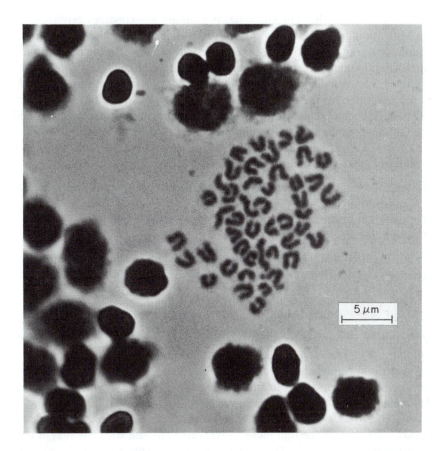

Fig. 7.6 Chromosomes of *Pleuronectes platessa*: the 'basic' fish complement of 2n = 48 rod-shaped elements.

sundry ways and all have large numbers of chromosomes and lots of DNA. The sturgeon are special in having many microchromosomes which assume importance in hybrid crosses.

Within the teleosts, Kirpichnikov identifies many parallels with the diploid tetraploid system (herrings v. salmon) discussed so fully by Ohno. Similar situations exist within the major order Cypriniformes and Cyprinodontiformes where individual families are characterized either by about 50 chromosomes or by about 100 with the Siluriform catfish probably showing the same evolutionary trend. The major groups not to show this tendency are the huge order Perciformes and the lesser Pleuronectiformes where the primitive 2n=48, or simple derivatives of it, seems always to hold. The postulated primitive chromosome complement for teleosts is 2n=48 with each of the 24 chromosomes acrocentric in form and of an evenly graded size – this is much the same as in modern pleuronectid fish (Fig. 7.6).

The dichotomy that chromosomal parsimony accompanies evolutionary sophistication and vice versa seems general. Whatever the evolutionary significance, the existence of close relatives with widely differing chromosome numbers presents problems and opportunities in hybridization.

Apart from its relevance to hybridization, very little of this absorbing tale of fish evolution on the grand scale is of value in fish breeding. Much more relevant is the finer scale variation in chromosome number and structure, and this can have significance to population studies and breeding programmes.

There are three main areas of intraspecific or intraindividual chromosomal variation:

1. variation in the number of chromosomes (2n) or the number of chromosome arms (NF);
2. variation in the banding patterns of individual chromosomes;
3. variation in chromosome content of the sexes.

The first two will be covered in this chapter, the question of sex chromosomes will be deferred to Chapter 8.

Intraspecific chromosome variation

Variation in chromosome number and structure is a major theme in fish cytology. The primitive teleost complement of 24 pairs of acrocentric or telocentric chromosomes of very even size range can by polyploidization and/or restructuring lead to most of the observed karyotypes today. The basic event in restructuring is believed to be 'Robertsonian' fusion whereby two acrocentric or telocentric chromosomes fuse in the region of the centromere to produce one metacentric chromosome with the loss of one centromere (Fig. 7.7(a)). This is equivalent to a translocation of one part of a chromosome onto another, which is a common event in studies of radiation-induced mutations and usually leads to quite gross genetic consequences involving duplication or deletion of genetic material. Another way in which a metacentric chromosome can arise is by the inversion of a segment of the chromosome which includes the centromere (pericentric inversion) (Fig. 7.7(b)). This too is encountered in mutagenesis work and it too has consequences other than altering the shape of the chromosome. By a combination of 'Robertsonian' fusion and pericentric inversion with or without polyploidization, almost any chromosome complement can be derived from the basic plan. Some authors also propose that centromere splitting could generate two telocentrics from one metacentric chromosome, but this would also require the generation *de novo* of two telomeres and there is no obvious way in which this could happen or any convincing argument that it does. Non-disjunction does generate additional centromeres and there seems every prospect that this mechanism could be implicated in chromosomal evolution. All of these chromosome changes are normally accompanied by serious adverse effects on fitness; the way in which this is overcome is not understood.

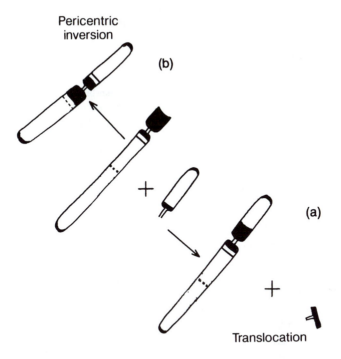

Pericentric
inversion

(b)

(a)

+

+

Translocation

Fig. 7.7 Chromosomal rearrangements. (a) A translocation involving two chromosomes. (b) an inversion including the centromere.

Variation in 2n values

Variations in chromosome complement can be seen not only between individuals of a species but also between different nuclei within an individual. This was clear from the early studies of fish cytology in *Salmo salar*. Svardson (1945) reported a modal chromosome number of 2n = 60 for this species with occasional nuclei containing 59 or 58 chromosomes. Svardson attributed this variation to non-disjunction although he accepted that there were no cases of 61 chromosomes. It is remarkable that the methods of that era were sufficiently reliable to rule out the artefactual 'loss' of chromosomes during slide preparation, but modern studies using far better methods all suggest a modal number of 58 chromosomes in Atlantic salmon (Roberts, 1970; Hartley and Horne, 1984a) and support the existence of variation within individuals. The situation is exemplified by the recent studies of Hartley (1987), Hartley and Horne (1984b) and Garcia-Vasquez *et al.* (1988) with wild stocks and farm stocks in Scotland and Spain. Table 7.2 summarizes the overall data.

In all of the stocks which were examined, the majority of fish had the modal number 58 but the Asturian stock had rather more of the variant types with 57, 59 or mixed chromosome numbers. The critical point is whether or not all chromosome spreads had the NF value of 74 which would suggest no difference of chromosomal material across the various karyotypes and would

Table 7.2 Chromosome number polymorphisms in Atlantic salmon. Data from Hartley and Horne (1984b); Garcia-Vazquez *et al.* (1988); Hartley (1987)

Chromosome numbers				Stock
56	57	58	59	
1	4(4)	25(19)	–	Scottish farm
–	1	21(2)	1	Norwegian farm
–	5	18(3)	1	Asturian farm
–	1	41(35)	–	Scottish wild
–	3(2)	23(18)	–	Scottish farm

Numbers in parentheses indicate individuals showing within-individual variation.

also confirm that the observations are real and do not represent imperfectly scored chromosome spreads. The numbers in the various studies were small, however, and too much should not be read into individual differences. The examination of 30 salmon of Scottish origin (Hartley and Horne, 1984b) showed two differences from the Spanish fish. First, the category with $2n = 56$ was observed, but secondly, within intra-individual variation, the NF value of 74 was generally found in association with the modal value of $2n$ whether it was 56, 57 or 58 whereas in those cells with chromosome numbers different from the normal value for the individual, the NF was not 74 but otherwise not specified. This must leave some doubt that intra-individual variation is homologous to interindividual variation.

The widespread occurrence of intra- and interindividual karyotypic variation in salmonids has often been referred to in reviews of fish cytology but very little attempt has been made to understand it structurally or genetically. Some attempts to define where Robertsonian fusions have occurred have been made by using banding techniques and will be touched on in the next section. The genetics remain an enigma in two important ways. First, the points at which chromosomes 'break' to rejoin in induced mutagenesis often have a deleterious effect; they may, for example, become recessive lethal mutations. However, it could be argued that Robertsonian fusion only involves heterochromatic regions around the centromere where genetic consequences per se are less likely than they are in euchromatin. The second enigma is that the meiotic mechanisms which occur in translocation heterozygotes often lead to semi-sterility. Figure 7.8 illustrates the way in which gametes with unbalanced chromosome sets can arise and such imbalance is normally very disadvantageous to a zygote.

Family studies with Atlantic salmon, to assess the inheritance of karyotypic variants, seem feasible but have not been performed. It seems likely that translocation heterozygotes would be semi-sterile. It is also possible that such effects would not occur in translocation homozygotes but instead, recessive genetic effects could become manifest. Whatever the outcome, it seems desirable from the fish farming point of view and for possible stock enhance-

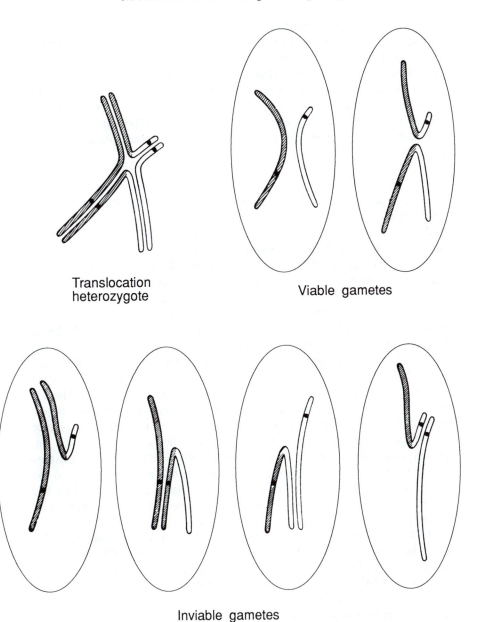

Translocation
heterozygote

Viable gametes

Inviable gametes

Fig. 7.8 The production of viable and unbalanced inviable gametes in a translocation heterozygote.

ment practice, to know more about the genetics of the karyotypic variation in Atlantic salmon.

Brown trout, although of much more diverse habitat than salmon, seem not to show much interindividual variation, having a modal number of 2n =

78 to 2n = 81. Rainbow trout show much greater diversity than salmon, with 2n ranging from 58 to 65 (Ohno *et al.*, 1965) and quite large amounts of intra- individual variation in some fish but not in others. There is a clear case for genetic studies to elucidate this complex situation.

Most of the examples of intraspecific chromosome number polymorphism came from studies with salmonids, but instances have also been reported in *Seriola* (Vitturi *et al.*, 1986), *Blennius* (Carbone *et al.*, 1987), and *Pleuronectes* and *Gadus* (Fan and Fox, 1991). It is clearly a widespread phenomenon in fish.

NF variation

Generally speaking NF values within the salmonids are species specific and stable; large scale inter- and intra- individual variation does not occur. Thus salmon in Europe have an NF value of 74, brown trout 101 or 102 and rainbow trout 104. The conservative inheritance of the major blocks of chromosomal material does seem to be normal although counting the arm number must be difficult with such large chromosome numbers, particularly when it is necessary to distinguish between acrocentric and telocentric chromosomes where different degrees of condensation can occur. There is one major example, however, of interracial variation which, if confirmed, should be relevant to any breeding programme. This is the apparent distinction between North American and European stocks of salmon (*S. salar*). The former has an NF of 72 whilst the latter is 74, and this distinction appears to embrace several putative races of anadromous and land-locked salmon (review, Hartley and Horne, 1984b). If substantiated, e.g. by identification of the 'transposed' material by banding techniques, this would represent the only single-factor distinction between these geographic races for which other diagnostics, e.g. protein electrophoresis, do not give individually definitive characteristics. The difference between the geographic races should also give pause to the thought of their admixture in farming but, in addition, stimulate the genetic investigation of the inheritance and consequences of such chromosomal polymorphism.

Salmon and trout have been exposed to a great deal of cytological study exceeding, perhaps, that on all other species put together. One overriding theme is that the salmonids are ancient tetraploids – first proposed by Svardson (1945) but developed into a credible scientific hypothesis by Ohno (1974). The extra set of genes and chromosomes in the newly evolved tetraploid can be regarded as a spare set which is no longer essential for the running of the organism and therefore free to mutate or change into something additional or useful. Thus, it is argued, the induction of tetraploidy in some distant salmonid ancestor laid the basis for widespread cytological evolution and even intra-individual variation within these fishes. On the other hand, the catastomids are also described as ancient tetraploids (Ferris, 1984) with a large chromosomal number of about 100, a DNA content to go with it but apparently no evidence of karyological variety.

7.4. BANDING POLYMORPHISM

Banding techniques and NOR staining are still at an undeveloped stage in fish cytology. Considerable interspecific differentiation by C-banding of centromeric and telomeric constitutive heterochromatin is readily apparent and should be of use in following the fate of chromosomes in interspecific hybrids. Some intraspecific variation for banding, particularly in relation to Robertsonian fusion, has also been described (Hartley and Horne, 1984b) but the genetics of this is unclear. Similarly, interspecific polymorphism for NOR numbers and locations is relatively common and intraspecific variation for Q-banding has been reported (Phillips and Zajicek, 1982). G-banding in fish is at an even more primitive state but Rivlin *et al.* (1985) and Gold *et al.* (1990) have made some progress with this approach and comment on other possibilities for future resolution of the fine structure of fish chromosomes.

7.5 LINKAGE AND CHROMOSOME MAPPING

That the number of gene loci is very much greater than the number of chromosomes in those organisms which were being studied at the turn of the century was one of the factors leading to the hypothesis that when two loci were linked genetically, in the sense that they did not segregate independently as predicted by Mendel's second law, they were located on the same chromosome. The frequency with which two alleles in a parent remained together in the offspring was a measure of the degree of linkage and since, by chance, two alleles on different chromosomes would segregate together in 50% of gametes anyway, linkage frequencies could be anything from 0.0 to 0.5. The final piece of theory which established linkage as a means of making chromosome maps was provided in 1911 by T.H. Morgan, who showed that for three linked loci, the linkage between one of the pairs was defined by the sum of the linkages between the other two pairs or their differences. In fact, it is the reciprocal of linkage, the frequency with which two loci separate or recombine during gamete formation, that is used for chromosome mapping. Thus for three loci the recombination frequency between A to C is the sum of that between A to B and B to C. Thus the chromosome map became depicted as a string of beads in which the beads were gene loci distributed linearly, albeit unevenly, along the length of the chromosome.

Cytologically, recombination became equated with the chiasmata which form during the first phase of meiosis. This comprised the process of crossing over between replicating chromatids and, as the centromeres moved apart during anaphase, the chiasmata progressively slipped down the chromatid bivalent as the threads unwound. This idea of a linear relationship between chromosome lengths and chiasma frequency needs reconsideration, perhaps, in the light of certain discoveries about crossing over and gynogenesis (p. 131).

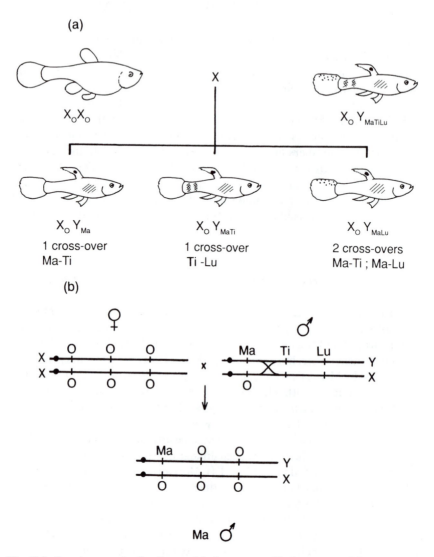

Fig. 7.9 Crossing over in the Guppy: (a) phenotypes; (b) chromosomal derivations.

Mapping chromosomes

It takes two gene loci to enable recombination lengths on chromosomes to be investigated by conventional means. In the fruit fly, *Drosophila*, with which the classical linkage work was done, there were hundreds of known multi-allelic loci and only four chromosomes. In fish, almost the opposite applies: relatively few loci are known and chromosomes are numerous but small. In the Guppy, however, many dominant colour pattern genes are known and these formed the basis of the first chromosome maps in fish. Data on the inheritance of 18 of these patterns were presented by Winge (1927, 1934) and used to construct a map of the sex chromosomes on which they were located. Winge's principal approach was to estimate the frequency with which genes crossed from the X to the Y chromosome and vice versa. Of the 18 genes, nine were never observed on the X chromosome and Winge suggested that they were alleles. Of the remaining nine, five were sufficiently well represented in the crosses to assess the frequency with which they moved from Y to X or vice versa. A typical cross is shown in Fig. 7.9(a). A female with no colour genes is crossed with a male in which the Y has three genes, *Maculatus* (Ma), *Tigrinus* (Ti) and *Luteus* (Lu). Amongst the offspring it is possible to distinguish cross-overs as those fish which lack any of the patterns. In Fig. 7.9(b), the chromosomal pattern of a single cross-over in the Ma–Ti region is illustrated. In this way, Winge built up a map of the chromosome in which Ma or one of the other obligate Y genes is closely linked to the sex determining part of the chromosome whilst the five genes *Coccineus* (Co), *Tigrinus* (Ti), *Luteus* (Lu), *Vitellinus* (Vi) and *Elongatus* (El) are progressively further from the Ma locus.

It is possible to look at these cross-over data not in terms of X and Y exchanges, but as cross-overs between each locus. Thus five cross-over possibilities can be identified and each of Winge's crosses can be interpreted in terms of inter-locus cross-overs. For example, cross 8 (1934), shown in Fig. 7.10, employs four loci, Co and Vi on the X and Ma and El on the Y, and the offspring phenotypes can therefore be scored for cross-overs as follows:

Ma El – no cross-over
Ma Co Vi – cross-over between Ma and Co – position 1
Ma Vi – cross-over between Co and Vi – positions 1-4
Ma – cross-over between Vi and El – position 5.

The actual numbers observed were 492 fish with no cross-over, 56 showing crossing-over in position 1–4, 2 in position 5 and 1 in position 1. The results of applying this procedure to all crosses are summarized in Table 7.3 from which it can be seen that the between-locus values range from 0.0 to 1.1% and it is possible to construct the map shown in Fig. 7.11. There is a problem,

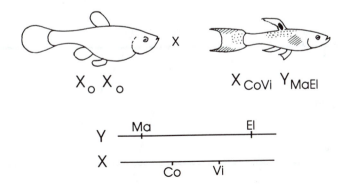

Fig. 7.10 Cross-over experiment in the Guppy.

Table 7.3 Reanalysis of cross-over data[*] from Winge (1934)

	Cross-over position				
	1	2	3	4	5
n	1412	93	1106	93	738
c.o.	6	1	8	–	3
%	0.4	1.1	0.7	0.0	0.4

	Cross-over region					
	1–2	1–4	1–5	2–4	3–5	4–5
n	611	282	440	1293	50	400
c.o.	4	12	16	76	–	–
%	0.7	4.3	3.6	5.9	0.0	0.0

Data are: n, number of possible observations; c.o number of observed cross-overs;
% c.o. × 100/n.

however, in that the estimates for crossing over for the larger regions are not
simply the sum of the estimates for the smaller regions. Thus the entire region
from position 1 to position 5 assessed as between-loci rates showed only 2.6%
crossing over whilst the small region 2–4 assessed overall was 5.9%. Similarly,
regions 1 and 2 taken separately comprise 1.5% crossing over but only 0.7%
if assessed jointly. In classical genetics it has long been known that one
cross-over exerts an influence on the probability of a second cross-over
occurring nearby, usually reducing it. This cross-over **interference** however
does not seem appropriate to explain the discrepancies in the Guppy data.

In addition to the non-additivity of cross-over frequencies on the Guppy
sex chromosomes, Winge (1934) also noted that crossing over of Vi or El,
respectively, within the sex chromosomes was dependent on the presence or
absence of the other gene. Clearly, the classical picture is not displayed by

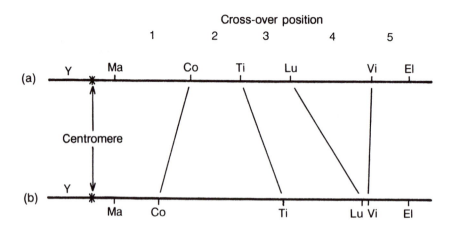

Fig. 7.11 The map of the Guppy Y chromosome: (a) after Winge, 1934; (b) data
recalculated.

these chromosomes and the possibility of sex chromosome specialization will be considered later (Chapter 8).

Unconventional mapping technique

The very small cross-over values seen in the Guppy were long taken to be typical of the small chromosomes seen in fish, but more recent work using unconventional means of measuring recombination has shown this not to be true. This novel means of measuring crossing over rates is gynogenesis. Full details of its application to fish breeding are given in Chapter 11, but for its implication for chromosome recombination, it is sufficient to note that in this form of parthenogenesis the egg is activated to begin the normal sequence of post-fertilization events without the involvement of a paternal chromosome set.

The second phase of meiosis is begun in gynogenesis but instead of proceeding to the establishment of a haploid female pronucleus, the egg is further activated to suppress the metaphase/anaphase stages of second meiosis so that the resultant genome is diploid. Diploid gynogenesis thus results in the retention within the egg of the sets of chromatids which arose during replication in the first meiotic division. This is the stage at which crossing over takes place. In the absence of crossing over, the gynogenetic offspring of a heterozygous female will be homozygous for one allele or the other. If crossing over occurs, however, between the centromere and the locus in question, the gynogenetic offspring will be heterozygous (Fig. 7.12). Thus the frequency of heterozygous gynogenetic offspring is a measure of crossing over between the centromere and the locus. One advantage of this method is that it requires only one locus and this is obviously important in fish, with few loci and many chromosomes. It should also be noted that there is no random association between centromere and locus, as between two loci, so that linkage frequencies are not limited to 50% but can range from nothing to 100%.

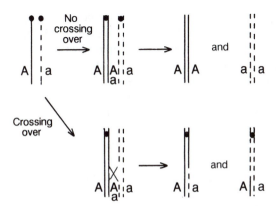

Fig. 7.12 Heterozygosity in diploid gynogenesis generated by crossing over.

The earlier studies on fish chromosome recombination by gynogenesis were conducted in carp by Cherfas (1977) and in plaice by Purdom *et al.* (1976). Much further work summarized in Chapter 12 has been documented since then, with a wide range of loci examined in many fish species. Recombination rates are very much higher than those observed in the Guppy which will be discussed in the next Chapter in relation to sex-chromosome specialization.

Recombination levels up to 100% have been observed by gynogenesis and this requires some explanation. One single cross-over results in recombination, but two cross-over events reinstates the original linkage, whether involving two or four strands of the conjugated chromosomes. With large map lengths the likelihood of two cross-overs should be significant. However, if this were so there should be an upper limit of 67% to gynogenetic recombination, and this clearly is not the case. The explanation may be that interference is high in fish chromosomes to the extent that only one chiasma can occur in any chromosome pair. A second anomaly is that the high overall cross-over rates implies that the small chromosomes of fish are particularly prone to recombination. Purdom (1984) reports that average crossing over in the plaice was 24%, implying a total chromosome map length of over 1000 units, which is not much less than that of the house mouse which has five times as much genetic material (DNA) as fish. There is a school of thought which believes that a chiasma in each chromosome pair is essential for the proper alignment of chromosomes on the metaphase plate of meiosis. With small fish chromosomes, this might imply that the classical view of linearity between cross-over relationships is not valid.

Hybridization and chromosome mapping

A further major development in the science of chromosome mapping in fish is the use of hybrids to generate heterozygosity. This approach was developed first in the sunfish hybrids (genus *Lepomis*) described by Wheat and Whitt (1973), who observed 15–22% recombination between 6-phosphogluconate dehydrogenase and α-glycerophosphate dehydrogenase in hybrid backcrosses. No linkage relationships were found for four other protein-coding loci. The major advance in mapping by this means, however, has been accomplished by using hybrids of the platies (genus *Xiphophorus*) by Morizot and his co-authors. Six distinct linkage groups were initially defined in these hybrids (Morizot and Siciliana, 1984), subsequently expanded to cover 14 multi-locus groups, and hence 14 chromosomes (Morizot *et al.*, 1991) with 3 further putative groups and a batch of unlinked loci adequate to mark the remaining 7 of the 24 pairs of chromosomes found in these fish.

Seventy-six loci were studied in these hybrid backcrosses. Such an array of allelic comparisons is unlikely to become available without hybridization, so generalization from this work takes on added significance. The major interpretation of the data, in this context, was that by comparison with the much sketchier data from other species, the platy linkages indicate the overall conservatism of fish linkage groups (Morizot *et al.*, 1991). In other words, the

Table 7.4 Linkage maps of 17 of the 24 chromosomes in *Xiphophorus* (after Morizot *et al. 1991*)

Chromosome	Loci and recombination frequency
i	Ada **21** G6-pd **20** Pgd
ii	Gpi-2 **10** Es-2 **6** Eno-2; Pk-2 **27** Es-5; Es-3 **33** Ldh-C **21** Mpi
iii	Me **6** Atp; Gapdh-1 **31** Guk-2 **33** Es-7
iv	Glo **44** Aat-2 **15** Gpi-1 **5** Pep-D; Pk-1
v	Mdh-2 **11** Glydh **36** Es-1; Es-4
vi	Np-2 **23** Tf **7** Amy; Guk-3 **7** Glns **28** Umpk
vii	Idh-2 **9** Galt-2
viii	Galt-1 **39** Hex
ix	Guk-1 **23** Pgm
x	Pvalb-2; Aco-2 **40** Pp-1
xi	Pgk **13** Pgam-1
xii	Gda **22** Pep-S
xiii	Pep-A **8** Tpi-1
U1	Aat-3 **35** Mdh-1
U2	Gdh **9** Pep-X
U3	Gapdh-2 **34** Pep-C
XXIV	Macro melanophore **13** Cah-l Sex chromosome

arrangement of genes within different fish taxa appears to have been changed little during evolution. The most surprising comparison, perhaps, was with similar data from salmonids (Wright et al., 1987), where chromosome doubling and much evolutionary development in terms of rearrangement and chromosome arm loss has been postulated.

The platy linkage groups are shown in Table 7.4. If confirmed, the constancy, or near constancy, of these within fish taxa will make the data of great value for fish breeding generally and for the identification of fish markers in fish gene manipulations.

Chapter eight

Sex determination

8.1 INTRODUCTION

Sex determination, the means by which the development of the differences between the sexes is initiated, is of fundamental importance in biology and particularly so for the study of genetics. In fish breeding it is also of obvious practical importance. The actual development of primary and secondary sexual characteristics is dependent upon hormonal cues, particularly those from the gonads themselves, but this leaves a circular argument and the fundamental question remains: what initiates the way in which the undifferentiated primordial germ cells are directed down one pathway or the other? How does an individual become a fully functional female or a male? The initial step is augmented or buffered by the hormonal sequences which follow, such that the sexual differences are normally clear cut, and blending in the form of intersexes is rare except as a transitional phase in those organisms that are sequential hermaphrodites, i.e. male or female at different stages of their lives.

The nature of sex determination has been the subject of much hypothesis, with complex arguments developing over the relevance of sex chromosomes and the balance between pro-male and pro-female tendencies in autosomes and sex chromosomes. Modern tendencies, however, are towards simpler systems, the ultimate is in mammals, possibly involving only a single gene locus which determines male- or female- initiating processes.

In fish, the study of sex determination is particularly interesting in view of the diversity of sexual forms and the presumed primitive or labile status of fish, in this respect, compared with the higher vertebrates, mammals and birds. The complexities of the various modes of sexuality in fish were discussed in Chapter 2. The range – from simultaneous hermaphrodites, through gonochorists lacking non-facultative secondary sexual characteristics, to species with marked sexual dimorphism and complex mating and parental care behaviour – must, however, imply an evolutionarily adaptive radiation of sex control, and this is in fact found to be the case.

The traditional way of assessing the hereditary pattern of sex determination in fish has been through the measurement of sex linkage, the extent to which inherited characters are not randomly associated with the sexes but are transmitted from male to male or female to female or in a regular criss-cross pattern from one sex to the other in successive generations. More recent developments involve direct attempts at assessing sex determination via the analysis of chromosome differences between the sexes, but the most modern trends are towards understanding genetic sex determination, either by following the nature of offspring derived from parents exposed to sex hormones to reverse normal sexuality, or by the use of special breeding techniques of chromosome engineering (Chapter 12). An immunological curiosity, the H–Y antigen, has also been used in some fish species.

8.2 SEX RATIO

Primary sex ratios in fish, i.e. the relative frequencies of females and males at hatching or as soon after the initiation of sex determination as possible, are usually 1:1 but secondarily may become very skewed. This may arise by the pursuit of different life styles of the two sexes. For example, in some anadromous brown trout, *Salmo trutta*, populations, only the females go to sea and the males remain in the river of their birth. Less extreme deviations occur by sampling error such as when males and females migrate separately to the spawning ground, as in marine flatfishes, or, more generally, by differential survival. Females frequently live longer than males and, in consequence, constitute increasing proportions of the older fish in stocks which spawn repeatedly.

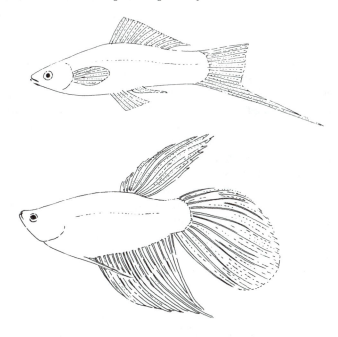

Fig. 8.1 Male of the swordtail (a) and the Siamese fighting fish (b): species with distinctive males but no clearcut sex determination mechanism (TL about 6 and 5 cm, respectively).

None of these secondary deviations have genetic significance in themselves, but the primary sex ratio of 1:1 is often assumed to arise because of a stable genetic mechanism controlling sex differentiation. The converse is more often stated, namely that skewed primary sex ratios, particularly where they are highly variable, indicate a lack of genetic control and the possible involvement of environmental cues. For hermaphrodite species this is quite easily accepted, because social and other factors usually guarantee an adequate representation of each of the sexes. For gonochorists, however, without sex reversal, such indeterminacy seems much less common.

The best documented cases of skewed and variable sex ratios in fish are for the swordtail, *Xiphophorus helleri* (Fig. 8.1) and the Siamese fighting fish, *Betta splendens*, both of aquarium fame. It has long been argued that the variable sex ratios in broods of swordtail offspring and the fact that sex reversal can take place – in this case from female to male – indicates a primitive state of sex control in this species, possibly reflecting polygenic control (Brieder, 1935; Kosswig, 1964). Thus the frequencies of females in the broods of these livebearing laboratory fish ranged from about 10% to 90% and, in addition, certain females tended to produce broods with fewer female offspring whilst others gave birth to broods in which females predominated – hence the implication of some sort of genetic control. Peters (1964) also records wide variations in sex ratio in fish derived from certain wild populations and further claims that the proportions of males and females was related to parental male size. Thus those male genotypes which were weakly masculinizing led to late maturation and therefore to large size, and a predisposition to sire females in the next generation.

Similarly deviant and variable sex ratios in several species of Xiphophorin fishes have been described by Kallman (1984) who casts some doubt, however, on the polygenic control hypothesis. These popular aquarium fish do not have sex-linked genes in the way that some other fish do, and Kallman points out that complex sex chromosome situations, as found in the platies, *X. maculatus*, can lead to variable sex ratios. In the next section we shall see how sex-linked genes permit a precise genetic interpretation of the behaviour of sex-controlling chromosomes.

The case of *Betta splendens* is rather different. This species (family Anabantidae) has many colour variants which are particularly brilliant in the male, but are not inherited differently by the sexes. Lucas (1968) showed that sex ratio varied widely, with the frequency of males ranging from zero to 97%. In addition, sex reversals were noted. Environmental variations in the form of temperature, water hardness and the age of parents, were all implicated in the variability of sex ratio but, unlike the situation in the swordtail, significant variation also occurred between replicate broods, and little or no genetic determination appears to be the case for this species of fish.

A radical study by Lowe and Larkin (1975) showed that after surgical removal of ovaries from mature females, almost half of the fish developed into males. Of these, about half sired broods of offspring when crossed with normal females. The sex ratio in these broods ranged from 50% females to 100%

females although the latter was the most common outcome. Lowe and Larkin discussed various genetic models and favoured the one involving polygenic sex determination.

There are several other species of the genus *Betta* but they are not so popular as the Siamese fighting fish and the sex determination patterns appear not to have been studied.

It may be difficult to establish the polygenic or multifactorial control of sex determination other than by the careful recording of sex ratios or by hybridization involving a related species of known sex chromosome type. Sex ratio studies are not well documented, but primary sex ratios in a recently described cyprinodont (*Ameca splendens*) with very low chromosome number are very variable (D.S. Woodhead, pers. comm.,1991) with rarely more than one male in broods of six or seven offspring. The critical studies in the context of polygenic sex control will be via selection and artificial sex reversal.

8.3 SEX LINKAGE

Since the differences between the sexes in fish are frequently seen as colour variations and since visual attributes such as colour are of immediate interest and appeal to amateur breeders or fanciers, it is not surprising that the first observations of sex-linked inheritance in fish were in connection with colour patterns. Indeed, most of our knowledge of sex-linked inheritance in fish concerns colour or melanin patterns and very few other characteristics display this sort of inheritance. The existence of sex linkage is usually taken to imply that there are sex chromosomes in fish: thus genes are described as on the X chromosome or the Y chromosome without evidence that either of these entities exists as distinct chromosomes. However, we shall deal first with the genetic aspects of sex determination and leave the cytology to later.

The first observed instance of sex-linked inheritance in fish was described by Johannes Schmidt in 1920; it concerned the Guppy, *Poecilia reticulata*. Two stocks of different origin both showed marked sexual dimorphism in that the males were small and brightly coloured whilst the females were larger and of a uniformly grey-green coloration. Furthermore, the colours of the males of the two stocks were quite different from each other and showed strict paternal inheritance. Whatever stock provided the female for a cross, the colour of the resultant male offspring was always more or less identical to that of the male parent. Schmidt reached no firm conclusion about the mode of inheritance of the colour patterns, but suggested that it was consistent with a sex chromosome hypothesis in which the female contained two X chromosomes, males one X and one Y chromosome and that the colour genes were located on the Y chromosome (Fig. 8.2). The Y-linked inheritance of colour patterns in male *P. reticulata* was firmly established by the work of O. Winge which was published in a series of classical papers on the genetics of the Guppy (Winge, 1922a,b, 1927) which was considered in some detail in Chapter 3. One important contribution of this work was the first demonstration in any

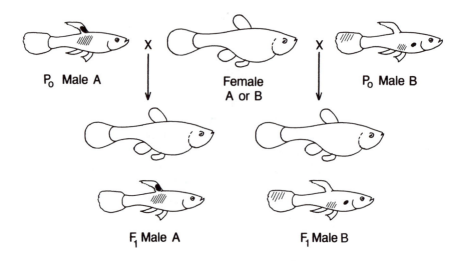

Fig. 8.2 Paternal inheritance in the Guppy described by Schmidt (1920). Male A is now recognized as Maculatus, Male B, Oculatus.

organism of paternally linked inheritance. Although described as the first instance of Y-linked inheritance, the distinction between X and Y has never been demonstrated cytologically in *Poecilia*, and homology with cytologically differentiated sex chromosomes in other vertebrates – or in invertebrates – cannot be assumed.

At the time of Schmidt's observations, and at the beginning of Winge's classical studies with the Guppy, sex-linked inheritance was also under study in two other fish species, namely the medaka, *Oryzias* (= *Aplocheilus*) *latipes*, and the platy, *Xiphophorus* (= *Platypoecilus*) *maculatus*. The medaka is not polymorphic for colour, nor does it show marked sexual dimorphism in nature – both sexes are a drab greenish colour. To demonstrate sex-linked inheritance, Aida (1921) used fancy stocks in which the normal female coloration was white with a variant red form in the males. When red males were crossed with white females, the offspring segregated into white females and red males. Thus the dominant allele for red (R) was carried on the supposed Y chromosome and r, the recessive allele, which does not produce red pigment was on the X chromosome. In the Guppy, coloration is part of the secondary sexual characteristic of the male. When females carry colour genes, they are not expressed. In other words colour in *Poecilia reticulata* is sex limited as well as sex linked. This is not the case with red in the medaka. If present in a female, the R allele produces a red fish. This genetic property has been extensively used to study induced sex reversal (to be detailed later) because its basic Y-linkage enables the genotypic sex to be assessed separately from phenotypic sex. Thus a female derived by reversal of sex determination in a genotypic male would show the red coloration if the Y chromosome carried the R allele.

Table 8.1 Summary of present conventions on sex chromosome symbolization

X	The basic symbol for a sex chromosome
Y	The convention now for a male determining homologue generating the XX female XY male convention
W	The convention now for a female determining homologue to X
Z	To replace X where the W symbol is used generating the WZ female ZZ male convention
O	The absence of a sex chromosome of any type

A third form of sex-linked inheritance is found in the platy, *Xiphophorus maculatus*, and has been widely studied by Gordon (1927), Kallman (1965) and many others. The first observations on this species were made by Bellamy (1922, 1924, 1928). Although platies are highly polymorphic for colour patterns, they lack the sexual dimorphism for colour exhibited so strikingly in the Guppy. Bellamy noted that a series of black spotting and red patterns could be identified in aquarium fish, namely White (W) a lack of large black pigment spots (wild type); Nigra (N) heavily pigmented; Pulchra (P) spotted and Rubra (R) red (p. 38). The patterns were only expressed singly in females but males could show two patterns. All attempts by breeders to get additional combinations were unsuccessful. By analysing crosses, Bellamy showed that the females transmitted only one pattern to their sons and none to daughters, whilst males with two characters transmitted either to either sex of offspring. Thus Bellamy concluded that the four characters were an allelomorphic series* and that the implied sex chromosome concept of the sexes was XX in the male and XY in the female (Fig. 8.3), the reverse of the case with the Guppy and the medaka. This was the situation for the aquarium fish as studied by Bellamy, but Gordon demonstrated later that an XX female, XY male situation existed in fish from the wild, and finally Kallman (1965) demonstrated that both mechanisms existed together in wild populations covering a major part of the natural range of *Xiphophorus maculatus*. There is thus a polymorphism in the sex chromosomes, or at least the linkage shown by them, within this species.

The symbols used for sex chromosome identification can be confusing because authors have used X and Y for heterogametic male and female situations. It is now conventional to use XX:XY for systems with male heterogamety and WZ:ZZ for female heterogamety. A simple summary is shown in Table 8.1. This does not entirely eliminate potential confusion where polymorphism exists as in the platy populations.

Kallman (1973) notes that there are other peculiarities to the platy sex determination mechanism. There is no evidence that separate populations ever existed such as could allow the separate evolution of different systems. The different sex chromosomes must therefore have evolved together. Kallman takes the view that there are three sex chromosomes in *X. maculatus* (X, Y and W) and that there is no need to postulate a Z to create ZW females as

*Later authors used different terminology and the anomalous red pattern, due to the presence of erythrophores, which are absent in the other phenotypes, is due to a different locus closely linked to the spotting gene (p. 41).

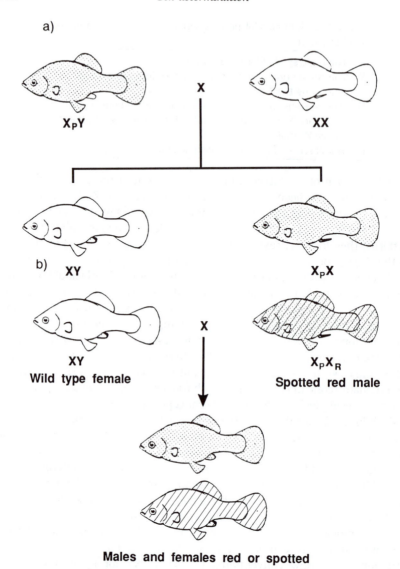

a)

X_PY XX

X

b)   XY X_PX

XY X_PX_R
Wild type female   **Spotted red male**

X

Males and females red or spotted

Fig. 8.3 The XX male, XY female system postulated by Bellamy (1922) for the platy,
Xiphophorus maculatus.

found, for example in birds. Thus the various sex control genotypes in the
platy are:

XY male – can be female
XX female ⎱
WX female ⎰
WY female ⎰ can be male
WW female ⎰
YY male – are always male

Table 8.2 Possible sex ratios in crosses within the platy system (female first)

		Normal		Atypical		
		XX	WX	WY	WW	XY
Normal	XY	1:1	3:1	1:1	1:0	1:3
	YY	0:1	1:1	1:1	1:0	0:1
	XX	1:0	1:0	1:1	1:0	1:1
Atypical	WX	1:0	1:0	3:1	1:0	3:1
	WY	1:1	3:1	3:1	1:0	1:1

Table 8.2 shows the sex ratios which would arise from some of the possible combinations of parents, and bearing in mind the lability of the sex determination process itself, almost any ratio can be expected.

Thus a range of sex ratios from all-male to all-female is possible in the natural environment – the question of why such a system is perpetuated when most other animals are characterized by a single system remains unanswered.

A further anomaly in the platy is that the W chromosome does not carry distinctive dominant colour genes as do the other two chromosomes. For this to be true, crossing over between the W and either of the other two within the region that controls sex cannot occur. This will be turned to next in connection with the organization of the sex chromosomes.

8.4 SEX CHROMOSOME MAPS

We saw in Chapter 7 that the sex-linked colour pattern genes of the Guppy were the first in fish to be used for the estimation of cross-over frequencies and the establishment of chromosome maps. The postulated X and Y chromosomes were distinguished by a number of genes. Winge (1927) originally believed that both X and Y carried genes which could not cross over to the other chromosome, but amended this concept in 1934 to one in which the Y chromosome has genes exclusive to itself whilst the X-linked genes could all cross over to the Y. The sex-differential chromosome, the Y, was thus characterized by a group of nine colour patterns which could not cross over and a further eight which could cross over to the X chromosome. In other words, the X and Y chromosomes were largely homologous but the Y included a sex-determining region. An eighteenth pattern (*Zebrinus*) was not sex linked but located on one of the autosomes.

Winge (1927) proposed that the obligatory Y-linked genes were, in fact, alleles of a locus which was very closely linked to the sex-determining part of the Y. It does seem surprising that the nine Y-linked colour patterns represent an allelic series whilst the remaining nine of the group studied by Winge are all at different loci but the facts remained that Y chromosomes always carried one of these nine genes and never more than one. The other eight sex-linked genes could be observed in combinations of more than two, and crossing over was observed between them as well as between each and the sex-determining part of the Y. Winge (1934) constructed the proposed map of the X and Y chromosomes in the Guppy as shown in Fig. 7.11(a). We have already seen in Chapter 7 that some anomalies existed in the partitioning of crossing over within the chromosomes but, overall, the conclusion that cross-over frequencies were low, in the range of 0% to 10%, remains valid. Low frequencies of recombination between sex-linked loci and the sex-determining region of the chromosome have also been described for the platy, *X. maculatus* (Bellamy and Queal, 1951; Kallman, 1965) and for the medaka, *Oryzias latipes* (Yamamoto, 1964).

This classical work on crossing over for sex-linked genes was generally accepted as indicative of the low recombination frequencies which might be expected in fish in view of the early cytological observations of large numbers of small chromosomes in these organisms. However, other studies on recombination, unrelated to sex determination, show that fish chromosomes do exhibit substantial cross-over map lengths. The first intimation of this came in studies of crossing over in gynogenesis, in which individual loci in fish exhibited recombination frequencies up to 100%. Details of this work are given in Chapters 7 and 12.

The second line of evidence for normal cross-over frequencies in fish comes from the more recent studies by Morizot and Siciliano (1984) and Morizot *et al.* (1991) on the segregation of protein polymorphisms generated by interspecific hybridization within the genus *Xiphophorus*. Of the 17 linkage groups which were defined by the analysis of pairwise segregation of 55 protein-coding loci, 12 showed map lengths exceeding 20% recombination. This, of course, is an underestimate of the recombinational length of chromosomes, because anything in excess of 50% means that the two loci concerned would segregate as if on different chromosomes. Of the 76 loci studied overall, only one was linked to the X chromosome, and showed a recombination value of 13% with the well-known sex-linked macromelanophore locus.

The conclusion therefore is that the close linkage of genes on sex chromosomes of certain aquarium species is atypical of chromosomes generally and may indicate the presence of a conserved chromosomal region for sex determination which is protected from disruption caused by crossing over. The existence of such a region, stabilized for example by an inversion, would be expected from classical studies to exert an influence on neighbouring regions of a chromosome. We shall see later that evidence for a morphological distinction between the sex chromosomes does exist but, unfortunately, not in those species for which genetic data are available. That chromosomal heterogeneity is not seen is inconclusive, however, since the condensed nature of metaphase chromosomes, the smallness of fish chromosomes anyway, and the present unrefined status of banding methods in fish all combine to make identification of small differences very difficult to achieve.

8.5 SEX REVERSAL AND SEX DETERMINATION

In Chapter 11 we shall look at the practical application of sex-ratio control in fish, probably the most important single genetic contribution to the practice of fish farming. The underlying scientific basis for this work, however, was achieved by the brilliant studies of a Japanese geneticist, Toki-o-Yamamoto, and the definitive review of artificial sex reversal in fish remains his authoritative account of his own work (Yamamoto, 1969b).

Various early studies on sex reversal had demonstrated that some features of sexual differentiation could be affected by the administration of sex steroids, particularly the display of secondary sexual characteristics (Grobstein, 1948), but no definitive alteration of functional sexuality was achieved. Yamamoto's successful studies were done on the medaka, *Oryzias latipes*, and employed two simple advances on anything that had gone before. First, he used a sex-linked colour gene so as to define the genotype with respect to sex, and secondly, he pursued his steroid treatments during the early phases of the fishes' life, before the fixation of the sexual differences became apparent.

The sex-linked marker used by Yamamoto was the gene for red body colour, described on page 32. In breeding, this was transmitted from father to son and, conversely, females were always white. Thus white defined the genotype of an individual as female, red the genotype of the male, whatever phenotypic sex was displayed.

Steroid hormones were administered to the fish in their food. Oestrogens or androgens were incorporated into fry diets and given to the fish from first feeding up until the fry reached a size (12 mm in length) at which the gonad is normally discernible as either male or female.

When a brood of medaka was fed with food containing androgen, all of the fish developed into males (Yamamoto, 1958), of which half were white and

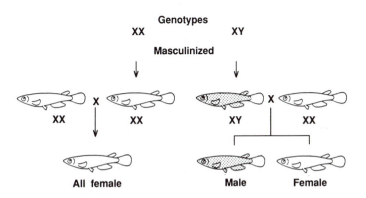

Fig. 8.4 Sex reversal by masculinization and its use to define the genetic sex control mechanism in the medaka, *Oryzias latipes*. Stippling donotes red.

half red. Yamamoto reasoned that the white males had the female genotype X_rX_r, and when crossed with normal females, also X_rX_r, they produced only female offspring, all of which were white, thus confirming his supposition. The red males (X_rY_R) were normal and bred normally (Fig. 8.4).

When broods of medaka were fed with diets containing oestrogens, (Yamamoto, 1953), all of the fish developed into females, but again half were white and half were red – confirming that red coloration due to the R gene, although sex-linked, was not a secondary sexual character as are the colour patterns in the Guppy.

When red females of the constitution X_rY_R were crossed with normal males the resulting offspring were expected to contain fish of the genotype YY (Fig. 8.5). These were identified by crossing with normal females. The offspring of these crosses were all males and, of course, red (Yamamoto, 1955).

It should at this point be noted that not all YY males in fish are viable – according to Winge and Ditlevson (1938), YY males of the Guppy, where the obligatory colour gene is homozygous, are non-viable whilst YY males with two different colour genes are normally viable.

The females of genetic constitution X_rY_R and the males X_rX_r therefore represented a reversal of sex from that which was defined genetically. Some authors dislike the term sex reversal since it implies the change of one sex into another, and propose the term sex inversion to cover these cases of male or female development as contrasted to the more genuine sex reversals which are required in hermaphroditism (Chapter 10).

Yamamoto attributed the success and repeatability of these sex reversal procedures to the fact that they were applied during a sexually indeterminate phase of the fishes' life. Work before and after Yamamoto's studies shows that the application of steroids starting later than this early phase can cause partial feminization or masculinization, but not fully functional sex reversal.

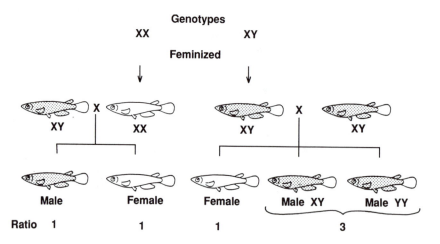

Fig. 8.5 Sex reversal by feminization and the production of YY males in the medaka.

according to Winge and Ditlevson (1938), YY males of the Guppy, where the obligatory colour gene is homozygous, are non-viable whilst YY males with two different colour genes are normally viable.

The females of genetic constitution X_rY_R and the males X_rX_r therefore represented a reversal of sex from that which was defined genetically. Some authors dislike the term sex reversal since it implies the change of one sex into another, and propose the term sex inversion to cover these cases of male or female development as contrasted to the more genuine sex reversals which are required in hermaphroditism (Chapter 10).

Yamamoto attributed the success and repeatability of these sex reversal procedures to the fact that they were applied during a sexually indeterminate phase of the fishes' life. Work before and after Yamamoto's studies shows that the application of steroids starting later than this early phase can cause partial feminization or masculinization, but not fully functional sex reversal.

The genetics of sex reversal has now been studied in several species of commercial importance and others of interest to aquarists. This applied work will be described in Chapter 11.

8.6 HYBRIDIZATION AND SEX DETERMINATION

The interest in the implication of hybridization for understanding genetic sex determination processes was stimulated in part by the famous dictum of Haldane (1922) that in F_1 hybrids with one sex absent, rare or sterile, that

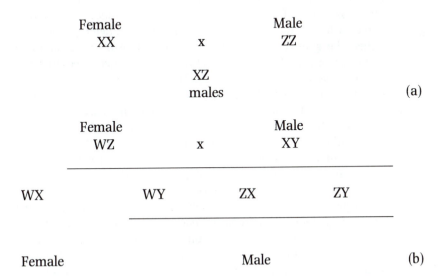

Fig. 8.6 The postulated consequences for sex ratio of crosses between tilapia species with differing sex determination mechanisms.

Regrettably, however, the initial hope of practical application was never realized simply because the hybrid broods always seemed to include a few females and these were able to start the undesirable cycle of reproduction and overpopulation.

The formal explanation for the predominance of male offspring in the tilapia hybrids was that two different sex determining mechanisms distinguished the two species (F.Y. Chen, 1969), one of the WZ female, ZZ male type, the other the more conventional XX female, XY male type. The predominantly male-only broods arise when the parents are both of the homogametic sex. Thus a female XX crossed with a male ZZ produces only XZ heterozygotes which, theory has it, will be male with a few exceptions (Fig. 8.6(a)). The other cross (WZ female and XY male) produces four zygotes of which three seem likely to be male, one female (Fig. 8.6(b)). This model has never been popular, and alternatives involving complex interactions of sex chromosomes and autosomal modifying genes (Hammerman and Avtalion, 1979) and outright polygenic inheritance (Mair *et al.*, 1987) have been postulated but not proved.

8.7 GYNOGENESIS AND SEX DETERMINATION

We turn to the final breeding tool of value in assessing the genetics of sex determination, namely gynogenesis. This form of parthenogenesis was touched on in relation to the estimation of recombination frequencies (p. 131) and will be dealt with in other contexts in Chapter 12. Its relevance to sex determination lies in the fact that gynogenetic offspring receive only maternal genetic material, and if the female is the homogametic sex then all gynogenomes will be female. Conversely, if the female is heterogametic (WZ or WY), then the gynogenomes will show segregation of the different sex chromosomes producing, for example, WW, WZ and ZZ genotypes in a ratio dependent on crossing over (Chapter 12). The WZ individuals produced by crossing over should be females and hence the sex ratio should be biased towards the female. The latter case was observed in plaice, *Pleuronectes platessa*, in one of the earlier studies of gynogenesis in fish (Purdom, 1969) and more recently in the zebrafish, *Brachydanio rerio*, under different circumstances by Streisinger *et al.* (1981). In neither case has other work confirmed female heterogamety in these two very different fishes. All-female production by gynogenesis is much the more common event. Stanley and Sneed (1974) describe it in grass carp, *Ctenopharyngodon idella*, Nagy *et al.* (1978) report female-only gynogenetic common carp, *Cyprinus carpio*, and several authors have confirmed the all-female nature of sundry gynogenetic salmonids.

The great value of gynogenesis for the elucidation of genetic sex determination modes, at least in a provisional way, is its speed – only one generation is required to produce a result and in fast-maturing species an answer may be available within weeks of hatching. A related procedure, androgenesis, has been exploited less in this context. It is more difficult than gynogenesis in practice since it requires the elimination of the chromosomes of the egg and their replacement by chromosomes from the spermatozoan, followed by diploidization. It would usefully distinguish between heterogamety and homogamety in the male.

Despite the apparent simplicity of the application of gynogenesis to analysis of sex determination, there have been some recent examples of complications. Avtalion and Don (1990) attempted to clarify the sex determination problem in tilapias using gynogenesis in *Oreochromis aureus*. The gynogenetic offspring of three normal females comprised both female and male individuals but the latter did not survive beyond 1 year of age and were therefore not used for breeding. Of the 13 F_1, gynogenetic females, four were used for a second generation of gynogenesis. Three of these F_1, females produced male and female offspring whilst the fourth produced only females. From a third generation of gynogenesis, using one F_2 female, female and male offspring were again produced.

Although the authors express reservations about the compatibility between these results and earlier chromosomal hypotheses for sex control in the tilapias, the explanation seems fairly straightforward. The original females were WZ, producing gynogenomes which were also predominantly WZ due to cross-over between the sex-determining part of the chromosome and the centromere – this type of crossing over, leading to high levels of heterozygosity, is not uncommon in gynogenesis.

Non-cross-over offspring would be infrequent but would include the geno-types WW and ZZ. The latter would be males, the former females which subsequently could only produce female gynogenetic offspring. The non-viable F_1 males could be explained as the consequence of using a highly inbred strain of *O. aureus*. The normal female is at least heterozygous for part of one chromosome and might be expected to be more viable than WW or ZZ. In *Oreochromis mossambicus*, a tilapia species believed to exhibit an XX female, XY male chromosomal sex-determining system, the gynogenetic offspring were always female (Pandian and Varadaraj, 1990). A gynogenetic confir-mation of the distinctive sex chromosome sets in two such closely related species is of special interest taxonomically and historically – these were the parents of the Malacca tilapia hybrids (p. 145).

The second, more complex case of atypical gynogenesis also concerns the appearance of male gynogenomes, but this time in a species for which firm evidence of XX female, XY male exists. The work began with the production of gynogenetic common carp of the mirror genotype (Komen and Richter, 1990). Diploidization was induced either during egg metaphase (ME) or at the first zygote mitosis (MI) – for details see Chapter 12. In offspring of the first group a low percentage (7%) of males and intersexes was found whilst after MI, almost 50% of gynogenomes were male or intersex. When a male of ME origin was crossed with a normal female, all offspring were female, confirming the absence of a Y chromosome in the male. However, when three females of MI origin were each crossed with normal males the sex ratios were very atypical. One female produced about 6% males or intersexes and the two others produced excesses of males and intersexes. Komen and Richter interpret this in terms of a hypothetical autosomal gene which in the homozygous condition induces maleness or intersexuality. This model neatly encompasses the sex ratios of the gynogenic crosses but unfortunately falls on consideration of the normal cross between the original female – which was postulated to be heterozygous for the allele – and a gynogenetic male which must be homozygous for the allele. Such a cross should, on the basis of the model, produce 50% male or intersex when in fact it produces only females. A further criticism would be that a sex modifier of the postulated sort would be expected to show up in normal crosses of this ubiquitous species. An alternative hypothesis might be that sexual developmental homoeostasis is disrupted by inbreeding and that the greater level of inbreeding implicit in MI causes greater deviation from normal sexuality (Chapter 12).

8.8 CYTOLOGY OF SEX CHROMOSOMES

Consideration of genetic sex determination so far has covered indirect, hereditary phenomena but the observation of sex chromosomes does permit a rapid, direct way to assess the situation. In addition, although the genetic evidence of sex linkage is a convincing demonstration of chromosomally determined sex, it does not adequately specify the extent of chromosome differentiation in this context. The complete specialization for sex determination of one member of a homologous pair of chromosomes in mammals and birds, and the parallel evolution of similar phenomena in insects (reviews, Ohno, 1967; Mittwoch, 1973) are of such basic importance that it has often been assumed that similar, if less completely evolved systems should also operate in lower vertebrates such as fish. On the other hand, the alternative view has often been stated that sexuality in fish is so primitive and labile that rigid control is unnecessary (Liem, 1966). Nevertheless, a considerable search has been made for cytological evidence of sex chromosomes in fish.

The manifestation of sex chromosomes

Cytologically, sex chromosome differentiation may be manifested in several ways. In mitotic material, the regular occurrence in one sex of one fewer chromosome than in the other sex, or the recognition of a dissimilar or heteromorphic pair by size or shape may be taken as evidence of heterogamety. In meiotic material, more precise evaluation is possible in view of the nature of chromosome pairing prior to metaphase of first meiosis. Thus the absence of a chromosome is indicated by the presence of a monovalent in each cell complement of chromosomes, and even slightly dissimilar pairs may be recognized when they are in close pairing even if not obvious in mitotic material. In addition, differential coiling or condensation of a chromosome or chromosome pair relative to the rest of the chromosomes may also be viewed as evidence of sex chromosome specialization. Examples of each of these phenomena have been recorded in fish.

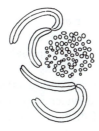

Microchromosomes

Large metacentrics

Fig. 8.7 Chromosomes of the fishes of the genus *Diretmus* (after Post, 1973).

In the older literature, the first evidence of cytologically distinguishable chromosomes as possible sex differentiators came from a study by Nogusa (1960) of the chromosomes of 64 species of fish of widely ranging taxonomic status. Nogusa examined sections of testis material and found evidence of sex chromosomes in three species. Surprisingly, one of these was a primitive chordate, amphioxus (*Branchiostoma belcheri*) in which pairing occurred between a large and a small chromosome in meiotic cells. This is surprising in that no sexual distinctions are apparent in the amphioxus-like agnathans. In the same work (Nogusa, 1960) no cytological evidence of dissimilar chromosomes was observed in a group of the more advanced agnathans, the lampreys.

The two other species for which Nogusa claimed evidence of sex chromosome differentiation were both teleosts of the order Percomorphi (sub-order Gobiodea). *Mogurnda obscura*, a sleeper-goby, showed meiotic pairing between a large and a small chromosome, both of which appeared to show earlier condensation than the remaining chromosomes of the cell during first meiotic prophase. *Cottis bairdii* did not have a pair of dissimilar chromosomes, but two elements showed precocious condensation as with *Mogurnda obscura*. Ohno (1974) has remarked on the curious similarity between the early condensation in presumptive sex chromosomes in these fish and the normal situation in meiosis of male mammals. A further example of early condensation of chromosomes is described by Post (1973) in fish species of the deep sea genus *Diretmus*. In one, *D. argentus*, the chromosome complement comprised 42–44 very small microchromosomes (less than 1 μm in diameter) and two enormous chromosomes of over 30 μm in length (Fig. 8.7). These latter chromosomes showed early condensation. In the second *Diretmus* species of unknown name, a similar situation occurred but with roughly double the number of chromosomes. Post concluded that the large elements might be sex chromosomes but corroborative evidence was lacking and only males were examined. The possibility that the unnamed species of *Diretmus* was a tetraploid also presented difficulties in that polyploidy and chromosomal sex determination are claimed to be incompatible (Orr, 1990). An alternative explanation of the function of the very large chromosomes might be that they were involved in the precise segregation of the microchromosomes, which must be difficult under conditions of normal meiosis. The appearance of chromosomes in *Diretmus* is unique but wider evidence of sex chromosome differentiation was from the study of other deep sea fishes.

The most definitive description of chromosomal specialization in relation to sex comes from the work of Chen and his colleagues (T.R. Chen, 1969; Ebeling and Chen, 1970), on a variety of deep sea and shallow sea teleosts. Of the deep sea species, 12 out of 25 showed evidence of male heterogamety whilst 5 out of 30 species of shallow water fish showed either male or female heterogamety. In much of the material, meiotic and mitotic evidence of heterogamety was available, including differences in size, abnormalities of pairing, and segregation of the different homologues into secondary spermatocytes.

Cytology of deep sea fish

Of the deep sea fish, the smelts of the family Bathylagidae (Salmoniformes) were the most consistent in their expression of heterogamety. Males were characterized by a large metacentric X chromosome which comprised 10% of the total chromatin, and a small Y chromosome. The latter varied among the four species examined. In *Bathylagus stilbuis* and *B. ochotensis*, the Y was assumed to be one of the larger acrocentric chromosomes, in B. wesethi it was the smallest acrocentric element and in B. milleri, the deepest living of the four species, the Y was one of the very small microchromosomes; Ebeling and Seltzer (1971) confirmed the homogamety of females for the large X chromosome in *B. milleri*. In another group of salmoniform fish, the Sternoptychidae, one of four species examined appeared to show male heterogamety. *Sternoptyx diaphana* has a diploid chromosome number of 35, and in the second metaphase of meiosis, either 18 or 17 chromosome elements were observed. The missing element was a large metacentric chromosome, which is unlikely to be overlooked, hence male heterogamety of the XO type was proposed. Chromosome analysis of females was not, however, reported.

Among the lantern fishes (Myctophiformes) examined by Chen, four cases of possible male heterogamety were found amongst 11 species examined. In two cases, *Lampanyctus ritteri* and *Parvilus ingens*, an XO male system was indicated by the presence of 47 chromosomes in the male, 48 in the female in the former and 49 chromosomes in the male for the latter. Females of *P. ingens* were not reported. In both cases the missing element was large, submetacentric and showed abnormal association during metaphase I and segregation at metaphase II of meiosis. In the other two lantern fishes in which males had distinctive chromosomes, *Scopelengys* sp. and *Symbolophorus californiensis*, abnormal pairing at meiosis comprised end-to-end alignment of two chromosomes at the periphery of the metaphase plate. In mitotic cells of *S. californiensis*, the pair could also be distinguished as submetacentrics of unequal size. In the fourth group of deep sea fish, the Melamphaidae, three out of five species, *Melamphaes parvus*, *Scopeloberyx robustus* and *Scopelogadus mizolepis bispinosus* had meiotic bivalents with end-to-end association in males. The individual elements in each case appeared roughly equal in length. A single species of the Apaplopasteridae, *A. cornuta*, showed no sex bivalent.

A further group of deep sea salmoniformes, the Galaxidae, was studied by Campos (1972). Of three species, one (*Galaxius platei*) showed male heterogamety with but a single large metacentric chromosome which was paired in the female. The Y was probably one of 11 telocentric chromosomes. The two other species, *G. maculatus* and *Brachygalaxias bollocki*, showed no evidence of sex chromosomes.

Excluding the two species described by Post (1973), out of a total of 47 species of deep sea fish, the frequency of chromosomal polymorphism associated with sex was 13 species out of 28 examined. Unfortunately, the nature of these species precludes their use in an experimental sense and confirmation of the sex-determining mechanism is therefore unlikely. There must, therefore, always be some doubt as to the significance of these diverse chromosomal polymorphisms. The question might be asked, for example, why these deep sea fish require what appears to be a sophisticated mechanism for sex determination. In none of the five families is there any marked sexual dimorphism. The Bathylagidae and Melamphidae are not themselves well known, but their close relatives show no dimorphism. The Sternoptichidae are not dimorphic, and only minor features distinguish sex in the Myctophidae (light organ arrangements) and Galaxiidae (ovipositor in female).

For the present there seems no prospect of ever breeding these deep water species, and their sex chromosome arrangements are of academic rather than applied interest. They do, however, indicate the sort of evolutionary mechanisms which might be observed if we look further than the very limited range of fish species so far studied by combined cytology and genetics. The Australasian galaxids represent obvious examples for further investigation.

Further inconsistencies in the interpretation of chromosomal polymorphism as sex determiners in fish have been discussed by Ohno (1974). Basically, the presence of dissimilar-sized chromosomes in one sex is not contentious and can be explained by the idea that sex chromosomes originate by the occurrence of a pericentric inversion (Ohno, 1967) in one of a pair. Thus exchange by crossing over is restricted and the way is open for evolution towards the hemizygous state in one sex as in birds or both sexes as in mammals. Precocious condensation and end-to-end pairing were also features of meiosis in deep sea fish and these mechanisms could also prevent crossing over. Ohno cites the observed crossing-over in sex chromosomes of *Poecilia* and other cyprinodonts against this concept, but as we have seen, crossing over is restricted in these fish and also in *Oryzias latipes*. In addition, in *Poecilia reticulata* a number of totally Y-linked patterns must either be allelic to each other, which seems unlikely, or part of a discrete section of the chromosome related to sex determination. Recessive lethal genes must also be totally linked to colour genes and sex determiners in *Poecilia* (Haskins *et al.*, 1970).

Dissimilar chromosome pairs and anomalous bivalents could still represent primitive sex chromosomes in which the two members are largely homologous, but the existence of apparent XO conditions are, in Ohno's view, a considerable enigma in that they imply an advanced evolutionary development in overcoming the problem of hemizygosity. Such sophistication among vertebrates appears restricted to birds and mammals. There are, however, grounds for believing that fish may not be as sensitive as higher organisms to the loss of chromosomal materials. Thus Robertsonian fusion (p. 122) is a commonly advanced explanation of variation in chromosome numbers, even in closely related species, and centromere loss is likely to involve also some loss of genetic material. Haploid fish can survive to beyond hatching (Purdom, 1969) and even to the onset of feeding and growth (Varadaraj, 1990), and grossly aneuploid cells may persist as mosaics and divide for some time.

The apparently diverse sex-determining mechanisms in deep sea fish therefore constitute a considerable paradox but they are of fundamental importance to the concepts of the evolution of sex determination. Progress via an experimental breeding analysis of such phenomena, however, must derive from studies with the more amenable shallow water species. Tissue culture genetics of the deep sea fish does, however, offer a realizable challenge for the future.

Sex chromosomes of shallow water fish

An extremely wide range of shallow water fish was studied cytologically by the older methods and much effort was expended in a search for sex chromosomes, mostly without success. Nogusa (1960) identified only three possible cases amongst 64 species including agnathans, elasmobranchs and 21 families of teleosts. Nogusa also reviewed the earlier cytological studies in fish, particularly those concerned with cyprinodonts for which genetic evidence of sex linkage was available. Thus Winge (1922a) and Vaupel (1929) were unable to demonstrate cytologically distinguishable sex chromosomes in the Guppy, and Friedman and Gordon (1934) and Ralston (1934) were similarly unable to find sex chromosomes in *Xiphophorus* species. Vaupel and Ralston both noted, however, that two large chromosomes showed atypical movement during meiosis in male fish and it was suggested that these might be sex chromosomes. Similar observations had been made earlier in *Umbra limi* by Foley (1926) and later by Bennington (1936) in *Betta splendens*. Nogusa threw doubt on the identification of these chromosomes as sex controllers and concluded that artefacts may explain their atypical behaviour. For *Betta splendens*, later claims that sex is not determined genetically (Lucas, 1968) seem to support Nogusa's views.

Heteromorphic chromosome pairs were reported in the eel, *Anguilla anguilla*, and the rudd, *Scardinius erythrophthalmus*, by Chiarelli *et al.* (1969). In both cases no attempt was made to relate the chromosome polymorphism to sex, but it was suggested that such a relationship might exist. These findings were made on cultured kidney cells and alternative explanations of the significance, or indeed queries as to whether the polymorphism is real, could be made. In addition, the eel seems an unlikely candidate for chromosomal sex determination. Sex is very indeterminate in eels and sex products cannot be identified until the fish are several years old. For a considerable further period, the gonads seem to contain both male and female germ cells and there is considerable evidence (reviewed by Bertin, 1956) that sex is finally determined by environmental factors. Nevertheless work using C-banding methods on European and American eels suggested heterochromatic differentiation of the ZW type in females (Park and Kang, 1979). On the other hand Sola *et al.* (1980) present convincing evidence of the absence of chromosomal differences between the sexes in the three species *Anguilla anguilla, A. rostrata* and *A. australis* representing the major world distribution of this genus. In fact the karyotypes show no differentiation between the species, which is not uncommon in related fish groups. The evidence would seem to be against the existence of differentiated sex chromosomes in the eels.

Further cytological studies have been made in shallow water teleosts in recent years but, again, the only definitive recording of sex chromosomes comes from the work of Chen and his colleagues (Chen and Ebeling, 1968, 1971; Chen and Reisman, 1970; Ebeling and Chen, 1970). By far the most spectacular heteromorphism occurs in *Gambusia affinis* (Fig. 8.8), a livebearing cyprinodont of world-wide fame for its use in the natural control of mosquitoes. In mitotic and meiotic cells, the females were distinguished by a single large metacentric chromosome which remained unpaired in meiosis. According to Chen and Ebeling (1968), this large metacentric chromosome may be regarded as the W within a WZ female, ZZ male format. In females a small acrocentric was proposed as the Z and, in males, two small acrocentrics showed end-to-end pairing in meiosis. Here at last, then, is evidence of sex

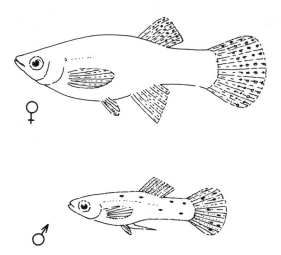

Fig. 8.8 *Gambusia affinis*: the mosquito fish, a species which is easy to breed, exhibits sex-limited inheritance and may have discernible sex chromosomes. Top, female; bottom, male; TL about 6 cm (female).

chromosome heteromorphy in an easily reared species (Fig. 8.8) showing marked sexual dimorphism; males are much smaller than females, have a gonopodium and show fiery aggression! In addition, the species is polymorphic for black pigment patterns in males and for black ventral spotting in females, a secondary sexual character of some importance in mating. In this species, then, the ingredients are available for a combined study of sex determination by genetic, cytological and experimental means.

Possible sex chromosome heteromorphy was also described in two further families of fish which occupy fresh or brackish water habitats. Chen (1971) described male heterogamety of the XY type in two among 20 killifish species of the genus *Fundulus*. *F. diaphanus* males had a metacentric Y chromosome which paired with a subtelocentric chromosome which was paired in females. In *F. parvipinnis*, the female complement resembled the *F. diaphanus* pattern but in males, the Y is presumed to be one of the smaller acrocentrics which number 47.

The subtelocentric elements observed in *F. parvipinnus* and *F. diaphanus* were marked by a constriction close to the centromere. Similar constrictions on acrocentric chromosomes in *F. heteroclitus* and *F. majalis* might indicate that these were homologous to the presumed X chromosomes of the heterogametic species, but in each case, males and females both had similar pairs of these acrocentrics. As with the related gambusiines, the genus *Fundulus* shows many examples of sexual dimorphism. *F. heteroclitus* is also a popular fish for experimental purposes, but no genetic systems other than protein polymorphisms are known, and the latter do not appear to show sex-linked inheritance in fish.

The third group of fish studied cytologically by Chen and Reisman (1970) was the family Gasterosteidae, the sticklebacks, which show seasonal sexual dimorphism for colour, and advanced levels of parental care associated with permanent morphological and behavioural sexual dimorphism. Two species of the five examined showed heteromorphic sex chromosomes, but of different types. *Gasterosteus wheatlandi* males were characterized by a single small acrocentric chromosome presumed to be the Y, while *Apeltes quadracus* females appeared to show heterogamety with a pair of chromosomes comprising a large metacentric and a smaller acrocentric element. An interesting feature of the biology of the sticklebacks is that the sexes segregate when mature. Males take up a solitary, territorial existence whilst females and juveniles school away from the males. There are therefore opportunities for quite different selection pressures in males and females and distinctive sex chromosomes may confer an advantage for divergence. A similar situation in *Poecilia reticulata* has been discussed in detail by Haskins *et al.* (1961).

Turning to studies with the more modern tissue culture techniques, evidence for sex chromosomes in salmonids was presented by Thorgaard (1977, 1983a). In rainbow trout, *Oncorhynchus mykiss*, from a range of natural habitats, males could be distinguished cytologically by the presence of a dissimilar pair of chromosomes involving a single subtelocentric which could not be observed in the female. The distinction was not always apparent and some natural populations displayed a greater prevalence of 'sex chromosomes' than others. Similar conclusions on rainbow trout were recorded by Chourrout and Happe (1986) and by Ueda and Ojima (1984a). In the latter case one strain of domesticated rainbow trout showed male heterogamety but another did not. Both strains had 30 pairs of chromosomes but no data were presented on NF numbers which might have indicated the presence of Robertsonian translocation. In kokanee (land-locked sockeye) salmon, *O. nerka*, Ueda and Ojima (1984b) suggest that a similar situation exists but where the chromosome number is also diagnostic, with females containing 58 chromosomes and males 57. No cytological evidence of sex chromosomes has been reported in the genus *Salmo* (Hartley, 1987) and it may therefore be the case that the observations in the genus *Oncorhynchus* represent a very early stage of sex chromosome differentiation in which a Robertsonian translocation by chance involves the male-determining 'gene'. Since these fishes can all be reared artificially and hybridized easily, there is obvious scope for further investigation.

An even more surprising application of modern methods is the demonstration by Haaf and Schmid (1984) of female heterogamety in the black molly, *Poecilia sphenops* var. *melanistica*. This popular aquarium fish is closely related to the Guppy and other livebearing cyprinodonts for which male heterogamety is indicated by past genetic studies but not by cytological study. Haaf and Schmid used C-banding and fluorescence techniques to identify heterochromatin in preparations from gill epithelial cells. Female heterogamety was shown as the presence of a highly heterochromatic chromosome amongst the 46 telocentrics found in both sexes. This the authors interpret as a W chromosome in a WZ female, ZZ male configuration. In addition, the heterochromatic region of the W and similar regions on two other chromosomes – probably nucleolar organizing regions – stained with fluorescent dyes in interphase nuclei such that a female nucleus could be distinguished by three fluorescing bodies, the male by only two. This appears to be the only example in fish of the 'Barr Body' situation common to mammals in relation to the inactive X chromosome of the female.

In a quite separate group of fishes, the Siluriformes, with over 24 000 species world-wide, evidence for and against sex chromosome differentiation exists. Agnèse *et al.* (1990) described cytological study of specimens of nine species of African catfishes of the family Mochidae. Female heterogamety involving a pair of metacentrics differing in size characterizes the seven species of the genus *Synodontis* for which female specimens were available. Other families, including the African Clariidae and Bagridae, were also reported to show female heteromorphy. In Asian catfish of the genera *Mystus* and *Platosus*, however, no sex chromosomes could be detected (Tripathy and Das, 1980).

Summary

To summarize the cytological evidence on chromosomal sex determination, several authors have recorded instances of chromosomal polymorphism which could be related to sex, but the most numerous and convincing cases derive from studies by T.R. Chen and his collaborators. Heterogamety appeared to be more common in deep sea fish than in species living at more moderate depths, but no explanation for this distinction is apparent. The deep sea fish in question show little sexual dimorphism. Amongst the less frequent cases of heterogamety in shallow water fish, several were associated with marked sexual dimorphism of morphological and behavioural types, but studies on species for which genetic evidence of heterogamety is well established failed to reveal cytologically distinguishable sex chromosomes. The most outstanding feature of these studies is the great variety of presumed chromosomal sex determination systems evolved in fish. Thus although mammals and birds appear each to be remarkably constant with respect to sex chromosome systems (Ohno, 1967), and similar conformity may be recognized in insects, variation within genera in fish is the norm and possibly occurs even within species. Conformity seems rare; only amongst the Bathylagidae were all species heterogametic and even here details distinguished the species. Thus it may be that sex chromosome mechanisms are very recent evolutionary developments in fish, and exhibit a sequence of parallel evolutionary trends with a common objective to isolate, in part, the genetic material which is relevant to the differences between the sexes.

8.9 AN ODDITY, THE H–Y ANTIGEN

This antigen was discovered during skin grafting experiments with inbred mice (Silvers and Wachtel, 1977). Because inbreeding leads to loss of heterozygosity, an established inbred line will comprise individuals which are genetically very similar to each other if not exactly identical in all respects. This means that the normal immune response, which occurs when tissue from one animal is transplanted into another, does not take place and such grafts are readily accepted. An exception to this was found in experiments with mice when male tissue was grafted into females. A weak graft rejection was observed, which has been attributed to the presence in the male of a specific cell-surface antigen associated with the Y chromosome and hence termed the H–Y antigen.

Immunological assays for the antigen can be based on the absorption principle. Silvers and Wachtel (1977) describe a cytotoxic and a cell-clumping assay in which an antiserum containing the H–Y antibody is first exposed to a test material and then assayed against mouse spermatozoa. If the test material contains H–Y antigen, the serum antibodies will be absorbed and will not affect the spermatozoa. If the spermatozoa are killed or agglutinated, the original test material is classified as H–Y negative.

The existence of H–Y antigen was quickly confirmed in other mammals, and then also in other groups of vertebrates including birds, where it was found in females, not males, thus illustrating not only its association with sex determination via heteromorphic chromosomes but also its probable antiquity in evolutionary terms. In fish, the antigen was found in males of the Guppy and platy (Muller and Wolf, 1979; Shalev and Huebner, 1980) and in medaka and tilapia, *Haplochromis burtoni* (Pechan *et al.*, 1979). These species, apart from *Haplochromis*, are known to show male XY genotypes. Study of H–Y antigen in species known to be characterized by heterogametic females have not been reported, but in the common eel, *Anguilla anguilla*, which is believed to have heterogametic females, Wiberg (1982) reports that females are H–Y positive.

The apparent ubiquity of the H–Y antigen led to a proposal that it was the 'single gene' responsible for testis development and therefore the simple agent of sex determination (Wachtel *et al.*, 1975). However, studies of sex reversal both ways in mice and some other pieces of evidence suggest that this simple interpretation is inadequate (review, Simpson, 1986). Nevertheless, even though it may not be the fundamental trigger of sex determination, does it still represent a marker for the heteromorphic sex? The work with fish suggests so far that this is broadly true and that an indication of heterogamety can be achieved by application of an H–Y antigen assay. That H–Y is not simple in fish, however, is suggested by the observation of it in the hermaphrodite species *Coris julis* (Duchac and Buhler, 1983) and *Anthias squamipinnis* (Pechan *et al.*, 1986). Whether this reflects an anomaly of the H–Y antigen or the secondary nature of hermaphroditism in the fish, i.e. residual sex chromosomes, remains to be evaluated. Shapiro (1988), for example, advanced the view that the genetic basis for H–Y antigen, the DNA itself, may be affected by the behavioural cues which induce sex reversal in these hermaphrodite species. Concomitant with both of these changes, he also reported changes in a type of satellite DNA (see Chapter 6).

Further development of the H–Y antigen story seems to be halted but it should be noted that the reaction itself has not been demonstrated in fish inbred lines. This may be because of the weakness of the immune response generally in teleosts but new developments with gynogenesis and sex determination offer attractive ways forward.

Chapter nine

Hybridization

9.1 INTRODUCTION

Fish hybridize relatively easily, and the variety of species crosses and even wider taxonomic combinations is vast. Lists of such hybrids occurring naturally and/or produced artificially have been published by Slaztenenko (1957), Schwartz (1972) and Dangel (1973), but the purpose of this chapter is to describe the types of study directed at fish hybrids and the use to which hybrids themselves are put. Some indication of the use of hybrids to generate genetic variance was given in Chapter 7 and the use of hybridization for assessing the genetics of sex determination was described in Chapter 8. Many other facets of the use of hybridization in fish breeding will be described, but first there is the fascinating subject of natural hybridization.

The importance of natural hybridization lies not so much in its positive implications for fish breeding – although as we shall see later the artificial production of hybrids is a powerful tool in fish culture – but more in its implications for evolutionary relationships. Hybridization in the natural environment is more relevant today as an environmental issue in the sense that introduced, or exotic, species might interbreed with and upset the natural genetics of native populations of fish. This problem is widely known as part of the trouble arising from the introduction of non-native species. Some ecological geneticists also feel that harm may occur where introgressive hybridization arises with, for example, domesticated strains of fish. The main thrust of the academic study of fish hybrids in nature, however, lies in evolutionary and taxonomic consideration. These aspects of natural hybridization were authoritatively reviewed by Hubbs (1955).

The much wider scope of artificial hybridization has not been reviewed so succinctly but Nicolyukin (1971) has emphasized the cytological basis for applied hybridization and Stebbins (1958) has comprehensively reviewed the entire field of hybrid inviability in living things.

Hybrids in the wild are usually recognized by their distinctive appearance and by the measurement of body proportions, i.e. morphometrics, or by counting fin rays or scales or teeth, i.e. meristics. These were the tools at Hubbs' disposal, but since his 1955 review electrophoresis and molecular methods of population assessment have become additional and very powerful

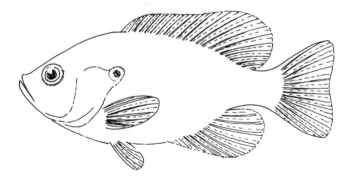

Fig. 9.1 *Lepomis cyanellus*: the green sunfish, one of the North American sunfish species
widely used in hybrid studies (TL about 20 cm).

tools. Protein differences between even closely related species are common; up
to a third of screened loci usually show divergence, and serve to pinpoint not
only the fact of first generation hybridization but also the possibility of later
generations, i.e. F_2, F_3..., backcrosses and even more distant crosses leading
to introgression. Mitochondrial DNA study provides an even more specific
probe in that it identifies maternal material and hence specifies the female
contribution to hybridization. These methods have been widely used in the
study of natural and artificial hybridization since Hubbs reviewed the evidence.

9.2 HYBRIDIZATION IN NATURE

The classical review of this subject by Carl Hubbs (1955), was based on his
lifetime study on hybridization within the sunfishes (genus *Lepomis*, Fig. 9.1)
and intimate knowledge of a range of other freshwater species in North
America. Most of the basic features of hybrids, as accepted today were covered
in this review, but hybrid inviability, the normal outcome of interspecific
crossbreeding, was not considered fully, possibly because it was not so obvious
in the sunfish crosses. The main characteristics covered by Hubbs were (i) the
taxonomic relatedness of putative hybrid parents, (ii) the intermediacy of
hybrids with respect to those features which distinguished the parents, (iii)
the enhanced performance of some hybrids – described as heterosis (now
sometimes euheterosis) but, more logically, luxuriance, and (iv) the sterility
which is often expressed by F_1 hybrids. We will consider some modern
examples of these phenomena and, indeed, the twin main theses of Hubbs'
review that hybridization of fishes in nature largely comes about through
environmental changes, particularly instability, or through imbalances between the species with or without the intervention of Man. A second major
theme which runs through much of the earlier work, and is of great

Table 9.1 Recent examples of hybridization of cyprinids in nature

Cross	Source
Scardinius erythrophthalmus × *Rutilus rutilus*	Thompson and Iliadou (1990)
Abramis brama × *Scardinius erythrophthalmus*	Economidis and Wheeler (1989)
Abramis brama × *Rutilus rutilus*	Economidis and Wheeler (1989)
Leuciscus cephalus × *Chalcoides chalcoides*	Economidis and Sinis (1988)
Leuciscus cephalus × *L. pleurobipunctatus*	Bianco (1987)
Alburnus alburnus × *Rutilus rubilio*	Crivelli and Dupont (1987)
Chondrostoma polylepis × *Rutilus arcasi*	Collares-Pereira and Coelho (1983)

importance still, is the problem of introgression, i.e. the extent to which F_1 hybrids can join in subsequent reproduction so as to 'mix' the genes of two species.

The group of fish in which natural hybridization seems commonest both in Europe and in North America is the family Cyprinidae. The many genera within this family comprise the species with popular names such as chubb, roach, minnows, dace and bream, all within the subfamily Cyprininae. Table 9.1 lists selected recent descriptions of natural hybridization within these fish. Some recent examples of natural hybridization within other families of teleosts are given in Table 9.2.

Taxonomic relationships

Comparison of Tables 9.1 and 9.2 will show that hybrids within the Cyprinidae frequently derive from parents in different genera whereas those from other families tend to be intrageneric. There is no clear explanation of this from taxonomic sources except that taxonomy within the Cyprinidae is probably more complex than that for other families and the placing of species in different genera less justifiable. On the whole, the intuitive message is that closely related species hybridize more readily than others, to which may be added the concept that species lacking specific individual courtship patterns are more likely to hybridize in nature than those species with specific courtship patterns (Raesly *et al.*, 1990). Many of the Cyprinidae exhibit mass spawning and some, such as the minnows, indulge in a mass whirling courtship activity which conceivably could

Table 9.2 Recent examples of natural hybridization in non-cyprinid fishes

Cross	Source
Salmo salar × *S. trutta*	Garcia de Leaniz and Verspoor (1989)
Salmo salar × *S. trutta*	Verspoor (1988)
Oncorhynchus tshawytscha × *O. kisutch*	Bartley *et al.*, (1990)
Oreochromis niloticus × *O.* spp.	Macaranus *et al.*, (1986)
Micropterus salmoides floridanus × *M. s. salmoides*	Maceina and Murphy (1989)
Solea aegyptica × *S. senegalensis*	She *et al.*, (1987)
Etheostoma zonale × *E. olmstedi*	Raesly *et al.*, (1990)

sweep any other susceptible fish along with it. Distant hybridization in nature is rare but we shall see examples of it in artificial hybridization.

Intermediacy

In most cases the natural hybrids are first observed as deviations from the known putative parents, and subsequent morphometric and meristic studies confirm the first impressions. The general conclusion that those characteristics in the hybrid are intermediate between those of the putative parents is amply confirmed, but intermediacy is not always exact. To measure this, various hybrid indices have been proposed. Hubbs and Kuronoma (1942) defined the most commonly used index:

$$V_h = (1/m) \sum_{i=1}^{m} \left[(X_{hi} - \mu_{1i})/(\mu_{2i} - \mu_{1i}) \right] \qquad (9.1)$$

where X_{hi} is the value of the ith of m characters in the hybrid and μ_{1i} and μ_{2i} are means of the ith character in each parent, respectively. Where intermediacy is exact, V_h will be 0.5 and any paternal or maternal tendency will be reflected in higher or lower values of V_h from 0 to 1. Where there is variation within the hybrids beyond the range shown by the parents, as for example with heterosis or luxuriance, the value of V_h can be greater than 1 or negative.

Table 9.3 Meristic and morphometric characteristics in the *Chondrostoma polylepis* and *Rutilus arcasi* hybrid system (After Collares-Pereira and Coelho, 1983)

	C.polylepis	Hybrid	R. arcasi	Hybrid index*
Meristics				
Scale no.				
Lateral line	68.8	49.1	42.6	75
Upper transverse row	11.6	8.6	7.8	79
Lower transverse row	5.3	3.9	3.5	78
Gill rakers	21.2	16.4	13.1	61
Pharygeal teeth	10	10	9.9	0
Fin ray no.				
Dorsal	8.0	7.4	7.0	60
Anal	8.1	7.8	7.2	33
Average of all meristics				55
Morphometrics				
Head depth/length	0.67	0.73	0.75	75
Caudal peduncle depth/head depth	0.63	0.64	0.62	NI
Caudal peduncle depth/total length	0.08	0.09	0.09	100
Average of all morphometrics				92

*NI, not intermediate.

Smith (1973) has criticized this approach on the grounds that observed differences can lead to subjective choice of character and proposes instead a method of principal components analysis (see Humphries *et al.*, 1981) for which graphical display can be used to highlight the levels of intermediacy and possibly the identity of generations other than the F_1.

In most of the examples in Table 9.1, the characteristics are intermediate, but not uniformly so. Thus Collares-Pereira and Coelho (1983), applying the Hubbs index to hybrids of *Chondrostoma polylepis* and *Rutilus arcasi*, found good intermediacy for meristic characters (0.55) but a distinct leaning towards *R. arcasi* for morphometric characters (Table 9.3). This was a case of frequent hybridization and although hybrids of F_2 or backcross origin could not be ruled out, it seemed unlikely that introgressive hybridization was involved in the lack of strict intermediacy. Similar bias towards one of the putative parents was found by Raesly *et al.* (1990) in their analysis by principal components methods of morphometric data from presumed hybrids in darters of the genus *Etheostoma*. Some of the departures from strict intermediacy may have been due to maternal effects, but this seems unlikely.

Hybrid vigour or luxuriance

We shall see later that this is a primary goal of artificial hybridization. In natural hybrids it is possible to detect vigour as over-dominance by the application of hybrid index analysis, but it is usual to find it by simple observation. Thus Hubbs (1955) cites the case of excessive numbers of sunfish hybrids in a stream to which they had migrated from the still-water pond in which they had hatched and where their vigour gave them territorial advantage over non-hybrids. In other cases superior growth rates can be shown. Crivelli and Dupont (1987) describe superior growth of hybrid bleak × roach (*Alburnus alburnus* × *Rutilus rubilio*) in a lake in Greece and list other references to this phenomenon. Similarly, Mir *et al.* (1988) showed that natural hybrids between *Barbus canis* and *Capoeta damascina* show morphometric values in excess of the means of both putative parents – meristics again were intermediate. Hybrid vigour could arise if full sterility occurred, i.e. if no gonad developed at all, such that all growth was somatic. Bianco (1982), however, failed to detect any difference between length at age for hybrids of bleak, *Alburnus albidus*, and chub, *Leuciscus cephalus*, in Italy even though both males and females developed only rudimentary gonads. This form of sterility is rarely seen, however, and the bleak × roach hybrids developed normal gonads.

An interesting possibility for luxuriance which might have practical application is that the hybrid between two species with different feeding habits might inherit the potential of both and thus widen its dietary options.

Hybrid sterility

The complete sterility implied by the absence of normal gonad development as seen in the bleak × chub hybrids reported by Bianco (1982) is uncommon, and the evidence normally advanced for or against hybrid sterility comes

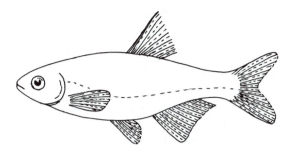

Fig. 9.2 *Alburnus bipunctatus*: a typical member of the bleak family of fishes (TL about 15 cm).

either from the presence or absence of F_2 or backcross individuals in a population, or from the experimental attempt to produce F_2 hybrids and backcrosses and even later generations. A special case is the presence of **introgression** which is the transmission into a species of genes from another species by hybridization. Many studies of natural hybridization address this issue, but meristic and morphometric methods, including discriminant function or principal components analyses, become increasingly inefficient at generations beyond the F_1. This is not the case with analyses by electrophoresis, where individual alleles can be traced with ease and where individual pairs of species may differ from one another at many electrophoretically detectable loci.

The normal hybrid situation in nature is that only F_1 hybrids are found; this is taken as evidence of hybrid sterility and a lack of introgressive hybridization. This is described for hybrids between the bleak (Fig. 9.2) and the roach in Greece (Crivelli and Dupont, 1987), white bass, *Morone chrysops* × white perch, *M. americana*, in the Great Lakes (Todd, 1986), salmon × brown trout hybrids in Newfoundland (Verspoor, 1988) and Spain (Garcia de Leaniz and Verspoor, 1989), and several species of sturgeons (Kozlov, 1970) in the former Soviet Union. The list is almost endless from many other crosses world-wide. The finding of fertile hybrids or their offspring are the exceptions. Thus Wood and Jordan (1987) claim that hybrids between *Rutilus rutilus* and *Abramis brama* were fertile in experimental crosses and that the F_2 and both reciprocal backcrosses with roach were fertile. The F_2 was indistinguishable from the F1 so that the observation of hybrids in a man-made reservoir could not be interpreted in terms of introgression. Similarly, many sturgeon hybrids are fertile when produced and tested in the laboratory (Nicholjukin, 1966), but evidence of introgression in natural populations is not so easily detected (Kozlov, 1970).

Introgression was claimed between a native cyprinodont and an introduced closely related species in Lake Balmorhea (Stevenson and Buchanan, 1973) (*Cyprinodon variegatus* × *C. elegans*). The circumstances were special. *C. elegans* is a riverine form carried into the lake into which *C. variegatus* was introduced. A hybrid zone was established between an irrigation flume and the lake shore with gradation towards the *variegatus* form between the hybrid zone and the lake. Both species could therefore be regarded as 'introduced'.

Several other cases of introgressive hybridization have been recorded. An occurrence not related to human activity was reported by Rakocinski (1980) in which sympatric populations of the cyprinids *Campostoma oligolepis* and *C. anomalum pullum* showed low levels of introgression on a morphological basis. Within salmonids the ubiquitous rainbow trout appeared to show introgression from hatchery stocks to natives in river and coastal populations (Campton and Johnston, 1985) as revealed by an electrophoretic analysis of isozymes. Many more cases are listed by Verspoor and Hammar (1991), but in the main the substantiation of introgression is often difficult and the general conclusion that it is uncommon still seems valid.

Environmental factors in hybridization

The most obvious environmental distinction in the field of fish hybridization is the paucity of natural hybrids in the sea by comparison with the wide variety in sundry fish families in temperate freshwater habitats. Hubbs made this point in his 1955 review and the situation remains so today. The familiar flatfish hybrids, plaice × flounder (*Pleuronectes platessa* × *Platichthys flesus*) (Pape, 1935; Von Ubisch, 1955), were the chief exceptions and explained by their mass spawning behaviour in association with differential migration of the sexes. The hypothesis was that when a 'shoal' of plaice females arrived at a site occupied by flounder males a hybrid swarm ensued. That it is this way round and not via the flounder female is demonstrated by the many artificial crosses which have been made and in which the cross with the flounder female does substantially less well than the plaice female, possibly because the plaice has much the larger egg size. Meristics and morphometrics define the F_1 hybrid, and no evidence of subsequent generations or introgression has been observed even though the hybrids are fertile in the laboratory.

Sundry other flatfish hybrids have been observed on occasions in the North Sea and there is evidence from chromosome studies of occasional hybrids within the family Centracanthidae in the Black Sea (Vasiliev *et al.*, 1984), but by comparison with freshwater species, marine examples of hybridization are still rare. The explanation by Hubbs that this is because of the greater stability of the marine environment remains valid despite the clear evidence of Man's impact on coastal regions and the growing intensity with which marine fisheries are exploited – perhaps more examples of marine fish hybridization can be expected in the future. In the meantime another exception to the rule is provided by Feddern (1968) who describes hybrids between two species of coral reef fishes, *Holacanthus isabelita* and *H. ciliaris* – the blue and green angelfishes (Fig. 9.3). The old adage that the exceptions prove the rule, where prove means test, is well met here in that these two species occupy different habitats in a shore/coral reef locality with an overlapping zone. This plus the fact that past glacial colonization is fairly recent provides the environmental

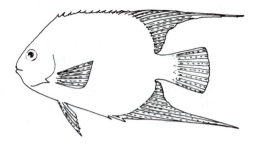

Fig. 9.3 *Holocanthus ciliaris*: the green angelfish, inhabitant of coral reefs (TL about 20 cm).

instability which more normally fits the freshwater scene than the marine. The hybridization itself points the way, perhaps, for future endeavours in the exotic fish breeders repertoire once the breeding problems with marine aquarium species are overcome.

Freshwater environmental changes such as the creation of new lakes or reservoirs, the construction of communication canals, and the general alteration of habitats including loss of spawning grounds have all been implicated in hybridization events since Hubbs' review and up to the present time. Many of the examples in Tables 9.1 and 9.2 illustrate this general theme, or the related one of non-native species introductions. Amongst the latter, the most spectacularly disastrous introductions have involved the popular game fish, *Salmo trutta* and *Oncorhynchus mykiss*, brown and rainbow trout, respectively. Introductions of these species were made to the eastern and central drainages of North America and hybridization with a wide range of cutthroat trout species, *Salmo clarkii*, and subspecies is reported to have been to their severe detriment (Ono *et al.*, 1983). To what extent it is hybridization per se which has caused this species hazard, as opposed to straight competition, is unclear. For example, Larson and Moore (1985) describe the simple encroachment by introduced rainbow trout and subsequent displacement of native brook trout in a Southern Appalachian Mountain stream complex. Similar loss of native species following introduction of game fish is occurring in Australasia (Pollord 1990), and here too, straight competition, rather than genetic contamination, appears to be the reason for the collapse of indigenous species.

Recognition of natural hybridization is in the nature of exception reporting and can be expected to achieve prominence beyond its logical significance. There is less cause for reporting the absence of hybridization or introgression even though on many occasions where one pair of species hybridize, and can be attributed to specific cause such as environmental change, other pairs of species are present but do not hybridize. Evidence for the sympatric existence of species without introgression has been presented for a few situations (e.g. Liley, 1966; Bartnick, 1970) but this does not reflect the overwhelming case that hybridization is the exception.

9.3 ARTIFICIAL HYBRIDIZATION

The interbreeding of different taxonomic groups has been a continuing saga in Man's domestication of animals and plants. Amongst animals, fish are very well suited to this approach and a vast literature exists on artificial hybridization, even greater than that we have already touched on with natural hybridization. Moreover, the scale and range of studies in artificial hybridization is very much wider than that for natural events. Thus crosses between mackerel and fundulus (Russell, 1939) represent an extreme in distant hybridization, equivalent to crossing a cat with a mouse in mammalian terms, whilst some of the interracial crosses in salmonids (Alm, 1955) and centrarchids (Philipp and Whitt, 1991) would barely be noticed anyway in the wild.

There are four basic ways of achieving artificial hybridization. The most natural way is the simple cohabitation in captivity of males of one species and females of the other. This is the usual way for aquarium species to be crossed, but pond or farm fish such as the tilapias can also be brought to successful spawning with this approach. In one sense, this method is very close to natural hybridization brought on following introduction of a new species into the habitat of another. A further development of natural cohabitation is to use hormone-stimulated sexual maturation: this can be achieved with pure hormones but is usually done by injection of the fish with extracts of pituitary gland, often on an empirical basis. The method is suitable for erratic or unreliable spawners such as grass carp, *Ctenopharyngodon idella*, and catfish, *Ictalurus* spp.

Artificial fertilization, whereby eggs and milt are hand stripped from the fish for mixture, is the most practical way of producing hybrids or, indeed, producing any genetic cross. It is widely practiced with many fish species and is particularly associated with salmonids. Finally, the fish may need stimulation by pituitary gland extract prior to hand stripping as, for example, with sturgeon which, because of their extensive migrations upstream, may be caught long before the natural period for ovulation and spermiogenesis.

Much of the effort put into the hybridizing of fish has been merely to satisfy curiosity. This is often overlaid with specific aims, such as to produce bigger or better fish, but usually with little prediction of what might happen. It is probably true to say that whenever two or more species are held together in captivity, they will sooner or later be used for hybridization.

More logical approaches to hybridization address specific problems such as the clarification of taxonomic relationships, the development of experimental tools for studying physiology, the production of genetic variance for selection programmes, the control of sex ratio (Chapter 8) and, finally, the production of superior fish for fish culture purposes.

Taxonomic studies

The artificial production of F_1 hybrids to check the validity of apparently new species in nature, is an old established procedure. Hubbs (1955) presents several examples and similar cases are still reported (Izyumov and

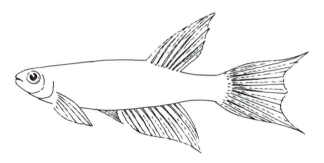

Fig. 9.4 *Aphyosemion calliurum*: a killifish species of aquarist acclaim (TL about 4 cm).

Gerasimenko, 1987; Burrough, 1981). A much less clear cut taxonomic aim is the establishment of relationships across a range of taxa by hybridization, based on the viability of the hybrids and on their phenotypic expressions of parental character. A huge range of crosses within the killifish (Fig. 9.4) group of genera has been produced by J.J. Scheel and mostly published informally in Killifish Letters in the 1950s and 1960s (Scheel, 1966).

Hybridization and developmental genetics

Hybridization as an experimental tool for the measurement of linkage relationships was discussed in Chapter 8. As a more broadly based methodology in fish developmental genetics, the studies on xiphophorin hybrids by Anders and his colleagues and on sunfish hybrids by Whitt and his colleagues are particularly outstanding, and although somewhat beyond the 'breeding' scope of this book, deserve brief mention.

The platyfish hybrids are especially valuable for the study of inherited neoplasms, particularly the melanomas which are amongst the commonest of fish tumours (Mawdesley-Thomas, 1971). From the earliest forays into hybridization between the two species *Xiphophorus maculatus* and *X. helleri* it became apparent that the normal pigmentation controlled by macromelanophore genes displayed enhanced expression (Haussler, 1928; Kosswig, 1929; Gordon, 1931). This seemed largely to be taken as an aspect of luxuriance in the hybrid, but subsequent wide-ranging studies (Anders *et al.*, 1984) indicate a complex polygenic system of regulator genes susceptible to selection and to X-irradiation enhancement to create massively invasive tumours of malignant form. The model may well enhance the view of hybridization as a priming agent for selection. Not only does it create variation for structural genes that code for proteins but also variation for the regulators which turn structural genes on and off. Similar genetic discord but of a more subtle form is illustrated by the work on hybrids of *Lepomis* sunfish (Whitt, 1981; Parker *et al.*, 1985). Heterogeneity of structural genes controlling the production of isozymes was achieved and, in addition, gene regulatory processes in these sunfish hybrids could be measured because of the genetic heterogeneity so produced. Thus it was possible to demonstrate the variation within time of expression of loci in embryological development and establish an understanding of the evolutionary divergence within this species range for both structural and regulatory genes. There appeared to be no relationship between the degree of divergence of species for structural genes on the one hand, expressed as genetic distances, and divergence for regulatory sequences for the switching on and off of this array, on the other.

Within the phenomenon of hybrid developmental anomalies these sunfish studies reveal the general incompatibility of regulatory systems within hybrids. Although in less dramatic style than that observed for the overweening malignancy of the platyfish melanomas, these observations generate a scientific basis for the general principle that hybrids are almost always disappointing to the would-be searcher for hybrid vigour but promising for the development of selection programmes.

Hybridization and selection

The development of melanomas in *Xiphophorus* species was one example of selection enhancement by hybridization albeit in a retrogressive way. Other developments represented by the vivid hues achieved in aquarium species could also be due to interspecific hybridization, although systematic study of such cases is not reported. The major colour morphs of goldfish and koi are certainly believed to relate to hybridization in their distant pasts, and homologous situations will be described later for hybridization within salmonids.

One purposeful study of enhanced variation in interspecific hybrids was the attempt by Tait (1970) to develop a trout hybrid by selection for use in the North American Great Lakes. The lake trout, *Salvelinus namaycush*, in these habitats had suffered severe depletion due to the invasion of the lakes by lampreys, which was facilitated by the opening up of navigation routes during the industrial revolution. By crossing a related species, the brook trout, *Salvelinus fontinalis*, with lake trout, it was hoped to produce a hybrid which by virtue of earlier maturity might reproduce before the combined effects of predation by lampreys and by Man reduced its numbers to inconsequential levels (Christie, 1960). Tait contributed to this wide-ranging programme by selecting hybrids with increased swimbladder function in the hope that it would increase their ability to penetrate the deeper waters, enabling them to reach the niches formerly occupied by the lake trout. These hybrids, known as 'splake' trout, and their parents were subjected to test in pressure vessels and considerable variance was found, with the lake trout most resistant and the hybrid and brook trout successively more affected by the pressure. Repeatability of the trait of pressure resistance was good but no reliable selection results were presented.

Similar aspirations to improve characteristics of Pacific salmon hybrids were outlined by Foerster (1968) in a general review of interspecific hybridization between five species of these fish. The hybrids themselves were unremarkable but some were fertile, and Foerster advocated a selection programme jointly to develop the flesh quality of sockeye salmon, *Oncorhynchus nerka*, with the early maturity of pink salmon, *O. gorbuscha*, or the rapid growth of chum salmon, *O. keta*.

The use of 'splake' trout in the Great Lakes to sustain fisheries is probably a better example of hybrid combination than selection. Certainly it seemed more resistant than the lake trout to the impact of lamprey predation, and with full fertility it was established as a self-perpetuating stock, at least in the hatchery. The use of selection following hybridization for specific traits as opposed to selection for general suitability has not been developed to any marked extent in commercial fish practice. The Soviet school of fish genetics is reputed to have developed a procedure for this using the common carp, *Cyprinus carpio*, × crucian carp, *Carassius carassius*, as the starting point. These intergeneric hybrids were not in themselves of much significance, showing low viability and very poor fertility. By patiently choosing the more fertile individuals and backcrossing to one or other of the parents, it was possible to develop fish which were predominantly of the form of one species but had specific features of the other, such as the prominent dorsal spine of the common carp or the deep body of the crucian.

Application of hybridization

The production, by hybridization, of new forms of fish which are more suited to specific needs is by far the most purposeful approach in this branch of fish breeding. In many cases, however, the aspiring breeder simply hopes that a hybrid will be better than the parents in general terms, i.e., in exhibiting hybrid vigour. This approach is very seldom rewarded. Most of the fish hybrids which have been produced in the last 100 years are significantly less fit than the parents and of little practical value. In a handful of species groups, however, significant progress has been made in the use of hybrids for practical fisheries purposes. This includes some cyprinids, bass of the genera *Micropterus* and *Morone*, the *Lepomis* sunfishes, the salmonids of almost ubiquitous deployment, and tilapias and sturgeons of more restricted use. Some of these developments will be reviewed briefly.

Bass hybrids

The bass are important sport and fisheries species in North America and, as elsewhere in developed countries for other species, overexploitation has led to the need to consider supplementation by the introduction of hatchery-reared stock. Widespread practice in the south-eastern United States is to stock with hybrids between *Morone saxatilis*, the striped bass (Fig. 9.5) common to coastal and freshwater habitats in the region, and *Morone chrysops*, a freshwater relative. The objectives of the cross were to exploit the faster early growth of the hybrid and its propensity to predate, and therefore control smaller prey species such as the gizzard shad, *Dorosoma cepedianum*. The early success of the venture in impoundments has been described by Bishop (1968) and, more recently, by Jahn *et al.* (1987) and in marine localities by

Fig. 9.5 *Morone saxatilis*: the striped bass popular with sport fishermen in the eastern USA (TL about 40 cm).

Yeager (1985). The superiority of the hybrid in these specific traits appears to be confirmed although a subsidiary aim to provide 'trophy' fish does not appear to have been met. In pond rearing too, the hybrid is reported to show good growth (Kerby *et al.*, 1987). These stockings have also led to the recovery of hybrids in commercial catches (Kerby *et al.*, 1987) as established by measurement of meristic and morphometric characters. These authors also report on the degree of sexual maturity of the specimens and predict the eventual appearance of F_2 and backcrosses in the wild – these predictions do not appear to have been substantiated and no evidence of genuine introgression has been advanced. Similar fears of introgression were advanced by Fries and Harvey (1989) concerning another bass hybrid, *Morone chrysops* × *M. mississippiensis*, in reservoirs in Texas into which both species but not hybrids had been introduced. Three extra-large specimens were found to be hybrids by the electrophoretic technique of isoelectric focusing. They were all F_1 hybrids, however, so did not represent introgression – they did provide evidence, however, of 'trophy' status for hybrids!

Hybrids of salmonids

Many different sorts of hybrids have been produced within the salmonids (Dangel, 1973). These are highly important food fish and even more significant supporters of sport fisheries, not only in the Northern Hemisphere, to which they are native, but also in Australasia to which they have successfully been translocated.

Artificial hybrids between the various species native to Europe were produced by Alm (1955) in one of the more comprehensive of the earlier studies. The species used were Atlantic salmon, *Salmo salar*, sea trout and brown trout, *Salmo trutta lacustris* and *S. t. fario*, respectively, Arctic charr, *Salvelinus alpinus*, and brook trout, *Salvelinus fontinalis*. Alm reached a number of conclusions.

1. Egg and alevin mortality was higher for the hybrids than for the straight crosses and could be related to chromosome imbalance and the relatedness of species.

2. Survival in later life was low for the hybrids.
3. Reciprocal hybrids were generally similar to each other unless there was great disparity between female size and male size.
4. Hybrid growth rate was mostly intermediate but sometimes greater than that of the parents.
5. Some hybrids differed very much in appearance from either parent.
6. Some hybrids showed some fertility but individuals in the F_2 were always less fit than in the F_1.
7. Most of the hybrids were of no potential practical value – the charr × brook trout was the most promising.
8. Successful hybridization in nature was unlikely.

It is interesting to note that the general compatibility of the *Salvelinus* species was reflected also in the splake trout hybrids discussed earlier. Similar results were obtained by Buss and Wright (1958) who included in the crosses the rainbow trout (then called *Salmo gairdneri*) and lake trout, *Salvelinus namaycush.* The *Salvelinus* crosses again came out best but an additional success was the 'tiger' trout produced by crossing female brown trout × male brook trout. This hybrid is so called because of its dramatically striped appearance (Fig. 9.6) which also contributed to its appeal to sport anglers. Production and sale of such fish was temporarily popular for put-and-take fisheries in Britain in the 1960s and 1970s but the fashion seems now to have passed. A similar obsession surrounded the possible use of rainbow trout × brown trout crosses

Fig. 9.6 The 'tiger' trout: a hybrid between *Salmo trutta* and *Salvelinus fontinalis* (TL about 35 cm). Courtesy of R.F. Lincoln.

and backcrosses. These were called 'brownbows' or 'sunbeams' but never achieved the popularity of the 'tiger'. The main reason for their lack of success was that they were not very viable, which is not surprising in view of the revision of the rainbow's taxonomic status and transfer to the genus *Oncorhynchus* as *O. mykiss* (Smith and Stearley, 1989).

The rainbow × brown trout hybridization programmes have now disappeared but for a time left a curious legacy in the British hatcheries that experimented with them. Because fish are mixed together, in the general commercial practice of grading and aggregating for size, the hybrids were mixed into the breeding pool and thus contributed to subsequent generations. This led to the appearance of trout of golden appearance. This was not the Mendelian characteristic discussed earlier, but a form which did not breed true but which represented some sort of hybrid pigment breakdown, possibly analogous to the phenotypic appearance of goldfish and koi which some also attribute to distant hybridization. The golden trout phenomenon seems to have petered out – possibly another case of failed introgression.

The annotated bibliography of interspecific hybridization of salmonidae by Dangel (1973) lists over 300 reports, but apart from the 'splake', none of the hybridizations seems to have generated lasting value in terms of practical fish breeding.

Hybrids of sturgeon

The sturgeons are important food species in Eastern Europe, the former Soviet Union and parts of Asia. Most of the species of commercial importance are large, have an anadromous life history and became sexually mature at a late age. These characteristics are not ideal for practical fish rearing with a closed life cycle, although the hatchery production of fry of several species for release into the natural environment has been a major feature of Soviet re-establishment of sturgeon fisheries. Nikolyukin (1971) reviewed the range of sturgeon hybrids to assess their potential for the artificial culture of sturgeon and concluded that a cross between the tiny freshwater sterlet, *Acipenser ruthenus*, (Fig. 9.7) and the mighty beluga or great sturgeon, *Huso huso*, which can reach a weight of over 1 tonne, offered the best prospects.

The hybrid produced by crossing the beluga female with sterlet male, called the *bester*, has since been developed for pond aquaculture in the former Soviet

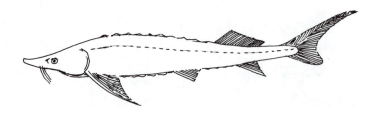

Fig. 9.7 *Acipenser ruthenus*: the sterlet, one of the smaller species amongst the mighty sturgeons (TL about 30 cm).

Union (Burtzev, 1972), and its main attributes were first, that it combined the growth rate of the larger parent with the freshwater tolerance of the smaller, and secondly, that it was fertile and could therefore be developed along the general lines of Soviet breeding practice. In a further development to achieve osmoregulatory tolerance in sturgeon hybrids, Burtzev *et al.* (1985) report on the cross between the Russian and Siberian sturgeons, *Acipenser guldenstadti* × *A. baeri.*

To return to the bester, several generations of line breeding have now been achieved and have established what appears to be an artificial breed specifically for fish farming use. Arefjev (1989) describes the chromosomal consequences of this line breeding. The original parents differed very little in chromosome complement, each having a large number of chromosomes (2n = 118) including many microchromosomes. There were differences in chromosome structure, however, and in the F_2 considerable variation with respect to the numbers of bi-armed and microchromosomes was observed. This variability decreased in the F_3 by the elimination of some microchromosomes and the apparent natural selection of the sterlet complement of bi-armed chromosomes. Arefjev concluded that a new, stable breed had been generated. If correct, this represents a remarkable genetic achievement which might have broader significance for the creation of improved breeds for aquaculture in other families of fish.

Other fish hybrids
Virtually all of the fish taxonomic groups that contain species which will breed in captivity provide examples of interspecific or intergeneric hybridization. In most cases the hybrids show intermediate inheritance, poor viability and sterility, and have limited practical potential. In some of the cyprinid crosses, however, this is not always true and future artificial production of such fish could well explore the benefits of hybridization.

Interracial hybridization

This was dealt with in part in Chapter 6 in relation to inbreeding and F_1 crossbreeding for performance and genetic uniformity. Further possibilities for intraspecific hybridization are (i) the use of different subspecies and (ii) the crossing of fish from distinct populations which have not, however, achieved the taxonomic status of subspecies.

The crossing of different subspecies is not well documented. One major example is the study by Philipp and Whitt (1991) of largemouth bass hybrids produced by crossing the northern and Florida subspecies, *Micropterus salmoides salmoides* × *M. s. floridanus.* The importance of this species as a sport fish in North America and its use in stocking programmes was mentioned earlier. Its popularity has led to a widening of the distribution of each of the subspecies themselves, to well beyond their natural range. This has brought them into contact with each other such that natural hybridization could occur. Philipp and Whitt were therefore concerned to assess the conservational impact of hybridization, not its purposeful deployment. The work was conducted in Central Illinois where winter temperatures are lower than those with which the Florida subspecies would normally have to contend. For growth rate and survival, the hybrids were intermediate between the parents, which themselves differed very significantly. The order of better growth and better survival was northern strain/hybrid with northern female/hybrid with Florida female/Florida strain. The authors concluded that the consequences of hybridization were potentially harmful.

The opposite conclusion on racial crosses was reached for the hybrid common carp, *Cyprinus carpio*, produced by crossing European and Chinese strains (Moav *et al.*, 1975). Three European 'domesticated' strains and one Chinese strain were compared with all possible hybrids. No data on survival were recorded but on growth rate, the authors found significant differences between all four strains and clear evidence of heterosis. A great deal of the variance for growth was attributed to environmental factors of pond and farm, and non-linear social interactions were taken into account in demonstrating the presence of strong heterosis. At face value, and bearing in mind the derived status of the weight-gain estimates, some of the hybrid comparisons looked more like intermediate patterns of inheritance, and only in one cross was heterosis marked. Marked heterosis is also claimed for intervarietal crosses of Chinese common carp (Zhang Jiansen, 1985), with hybrids showing 20% greater growth than the better parent, and sustained improvement was also seen in F_2 and backcrosses. No statement was made on the degree of inbreeding of these stocks, but the inbreeding history of the European stocks was known and is commensurate with a low level of heterosis.

Very similar results were obtained by Ayles and Baker (1983) with various domesticated strains of rainbow trout reared in prairie pot hole ponds in Manitoba. Seven strains and their crosses were assessed in a range of pot holes. On an analysis of variance on growth and survival for the pure lines, environment ranked the greatest component, strain was next but strain/environment interaction was low. The component of variance for the crossbreds was of the same rank but more extreme. Some crosses appeared to show heterosis for growth.

That some hybrid vigour occurs more often for interracial crosses than for interspecific hybrids seems reasonably well established. For domesticated strains where there is some evidence of inbreeding there may also be some evidence of heterosis, but in neither case does it appear as if strong heterotic forces are at work.

9.4 SUMMARY OF APPLIED USE OF HYBRIDS

The conclusions reached by Hubbs (1955) in his review of hybridization remain largely valid but with some qualification. In the majority of cases, hybrids are intermediate for those characteristics which distinguish the parents but a bias towards one parent or the other is very common. This non-additivity of inheritance is compatible with the concepts of developmental homoeostasis (Lerner, 1954) and with the seemingly low levels of additive genetic variance for metrical traits in fish. Hybrid vigour is not uncommon, but appears frequently only in crosses of closely related species or subspecies. Heterosis is found in some racial crosses but more often in crosses of domesticated stocks where past history indicates that some inbreeding has taken place. In terms of fish breeding, the most purposeful developments have used the intermediacy phenomenon rather than hybrid vigour so as to combine two or more desirable traits. The best examples are the splake trout, combining factors which assist it to resist lamprey predation, and the bester sturgeon, which combines the growth rate of one species with the osmoregulatory facility of the other.

Hybrids are usually of little practical value. Most show some level of inviability which is generally greater for the wider taxonomic crosses. Hybrid sterility is widespread but rarely takes the form of degenerate gonads. Gametogenesis may be normal, but unbalanced chromosome sets confer sterility, and on those occasions where normal gametes are formed and can generate viable zygotes in artificial fertilization, there appear to be behavioural or other barriers to realized fertility in nature. Natural hybrids are predominantly of the F_1 generation. Later generations and introgression are rare, even with interspecific crosses.

Hybridization has an important place in fish breeding, used intelligently.

9.5 HYBRIDIZATION AND INTROGRESSION

A major concern of environmentalists and conservationists today is the prospect that the accidental release of non-native forms of a species, or the deliberate introduction of such fish for stock augmentation purposes, could lead to lasting damage to the genetic integrity of the native stock. The fear is normally expressed either in terms of the disruption of 'co-adapted gene pools' by the introduction of alleles not already part of them or by the lowering of fitness overall by the introduction of inbred and therefore weak stock or, of course, by both events acting together.

There is ample evidence from electrophoretic and other studies that species can exist as genetically distinct subpopulations. Where geographic isolation occurs it seems almost universal that different genotypic frequencies will arise, and a good deal of controversy exists as to whether this is adaptive radiation or purposeless drift. Whatever the origin, the divergence is such that populations almost always comprise different mixtures of the same set of alleles. Occasions where a population difference can be expressed as the presence or absence of an allele are extremely rare, if they exist at all.

Are the mixtures of alleles, the so-called gene pools, really co-adapted? This term was coined by classical geneticists working with species such as *Drosophila*, the fruit fly, which lives in multitudes of environmental niches, has small chromosome numbers and numerous chromosomal structures which inhibit or prevent crossing over. None of these attributes applies to fish generally although some might apply specifically, such as niche choice for certain endangered species and the existence of reproductive barriers for seemingly sympatric populations. For most fish, and certainly for those of commercial importance, habitats are often wide and confluent, chromosome numbers are large and mechanisms to suppress crossing over mostly absent. The need for highly conserved co-adapted gene pools and the mechanisms for promoting them therefore do not seem to exist in most cases. Indeed, if it is accepted that a reproductive barrier will generate population divergence, the corollary is that the genotypes in the undivided state represent a compromise, not a co-adaptation.

A further unchallenged precept of population genetic studies over the past 20 years is the presence of considerable genetic heterogeneity in populations – in other words, lots of alternate alleles. If we take the case of a species with 48 chromosomes, the basic set in fish, and one locus on each with distinctive alleles – an unrealistically low level – the number of different genotypes comprising mixtures of the two homozygotes and the heterozygote for each of the 24 chromosome pairs is 3^{24} or approximately 280 billion (2.8×10^{11}) which is far larger than any conceivable fish population. Given that each chromosome contains many loci with alternate alleles, the chance of any two organisms being identical following sexual reproduction is negligible.

It follows then that much of the potential variability of individuals within populations has never been realized and that evolution has not tried out all of the possible combinations of genotypes. The notion of a unique solution to environmental constraints in the form of a co-adapted gene pool is, therefore, not valid. A further consideration is that as with the genetic code itself, the genotypic/phenotypic code is probably degenerate to the extent that a phenotypic end result can be achieved by a very large, perhaps almost infinite, array of genotypes. This is a corollary of the concept of developmental homoeostasis.

Populations are resilient provided that opportunity exists for sexual reproduction within a range of natural environments. Much has been written of endangered species – Ono *et al.* (1983) list over 100 in North America alone – but in the vast majority of cases the threats are environmental, not genetic. Some instances are attributable to competition from introduced alien species, but for this to be mediated through genetic agencies is the exception. Some of the decline in distribution of cutthroat trout has been equated with hybridization with introduced trouts. Outside this, very little evidence exists of 'genetic pollution'. It is the inviability of the hybrid or its sterility which gives rise to the fear of undesirable impacts of introgression, but it is also a fact that these phenomena drive the process of natural selection which reverses any adverse effect, given time.

The primary threats to fish species throughout the world are represented by straightforward competition with more vigorous introduced species or by destruction of habitats, particularly those essential for reproduction and recruitment. Genetic threats per se should be much further down the priority lists than these essentially environmental issues.

Conservationists often quote the words of Ernst Mayr. In *Animal Species and Evolution*, he writes "... genes of a gene pool are the product of a long history of selection for optimal interaction, one would always expect to find a drastic loss of fitness in interpopulation crosses. Surprisingly, the exact opposite is sometimes the case." (Mayr, 1963).

Chapter ten

Atypical modes of sexuality

10.1 INTRODUCTION

Gonochorism, the existence of separate sexes, is the normal mode of sexuality in animals, but a range of other possibilities exists naturally or can be induced artificially and fish, more than all other vertebrates, display a remarkable facility for such developments. Two major categories of atypical sexuality can be discerned: **hermaphroditism**, where an individual can be male and female either simultaneously or successively during its lifetime, and the state of **unisexuality**, i.e. species represented only by females, where sexual reproduction normally depends upon males of other species. Unisexual species thus depend on hybridization for their perpetuation and possibly also for their origin in the first place. These two phenomena, hermaphroditism and the existence of female-only species, are not intimately related but they do come together in one fish species, *Rivulus marmoratus*, which will be discussed more fully later, and also provide a link between the subject of gonochoristic sex determination as discussed in Chapter 5, hybridization covered in Chapter 9, and sex-ratio control, the applied side of sexuality, to be described in Chapter 11.

10.2 NORMAL HERMAPHRODITISM

Normal hermaphroditism is strictly defined as the functional expression of bisexuality as the normal mode of reproduction for a species. It is thus a characteristic of the species rather than of the individual. Those cases where an occasional fish shows signs of both sets of primary sexual characteristics, despite a normally gonochoristic life style for the species as a whole, are regarded as abnormal hermaphrodites. Their significance will be reviewed in Section 10.3.

Hermaphroditism in a species implies, of course, that genetics can play little or no part in sex determination as observed in gonochorists. The subject could therefore be regarded as having little to do with genetics and fish breeding. However, there are cues which initiate the necessary switch from one sexual form to the other, and not all individuals in a hermaphrodite species undergo sex reversal. So genetic influence on this process is still possible but not, it would appear, yet subject to critical study. Of more importance in terms of breeding, some hermaphrodite species such as *Sparus auratus* are being developed as farm fish; in addition, there is a growing amateur hobby of breeding tropical marine aquarium fish, and such species are much more likely to include hermaphrodites than the more traditional freshwater aquarium favourites. It is a significant feature of hermaphroditism in fish that most of the species exhibiting it are marine, tropical and basically reef-loving. It is therefore necessary to review briefly the nature of natural hermaphroditism.

The subject has been reviewed comprehensively by Atz (1964), Reinboth (1970), and in relation to modes of sex reversal by Reinboth (1988). Three basic forms of hermaphroditism are found in fishes:

1. **protogynous** hermaphroditism, in which individuals develop first into females and turn later into males – this is the commonest form;
2. **protandrous** hermaphroditism, in which the male state differentiates first;
3. **synchronous** hermaphroditism where both male and female states coexist functionally – this is the most uncommon form of hermaphroditism.

Curiously, a fourth possible pattern, **cyclical** hermaphroditism, in which sexuality can oscillate, does not occur in fish even though it is not uncommon in invertebrates. The sex reversal which occurs in protogynous and protandrous hermaphroditism is permanent.

There is a clear evolutionary pattern to hermaphroditism in that although the phenomenon has a wide taxonomical distribution, certain taxa are represented within the catalogue of the various forms of hermaphroditism very much more commonly than others (Table 10.1). Thus most of the fish groups exhibiting hermaphroditism occur within the order Perciformes – that group of fishes accepted as evolutionarily the most advanced and expressing wide adaptive radiation. So the hermaphrodite state is not to be regarded as primitive, and all reviewers, in fact, assert that evolution towards it and back again towards gonochorism are modern developments. In this context it is of interest that individual families of the order Perciformes exhibit varying forms of hermaphroditism and also gonochorism. The sparid family probably represents the most diverse group in this context, containing examples of protogynous, protandrous and synchronous hermaphrodites as well as straightforward gonochorists and others suggesting recent change from hermaphroditism.

The biggest group of Perciform fishes to exhibit hermaphroditism is the family Serranidae, which includes species of great abundance and of commercial fishery importance. As Atz (1964) points out, the type of sexuality runs within genera – thus *Serranus* itself contains many synchronous species whilst *Epinephalus* are protogynous hermaphrodites and the sea basses *Roccus* and *Dicentrarchus* are gonochorists. The Sparidae, however, include genera such as *Boops* and *Pagellus* with species exhibiting differing forms of

Table 10.1 Various forms of hermaphroditism amongst taxonomic groups of fish

Order and family	Genus or species	Form of hermaphroditism
Myctophiformes		
Four families		All synchronous
Synbranchiformes		
Monopteridae	*Monopterus alba*	Protogynous
Cyprinodontiformes		
Cyprinodontidae	*Rivulus* spp	Many gonochorists
		one synchronous
Perciformes		
Serranidae	*Serranus* spp.	All synchronous
	Epinephalus spp.	All protogynous
Sparidae	*Sparus* spp.	All protandrous
	Boops spp.	Protandrous or gonochorist
	Pagellus spp.	Protandrous, protogynous or gonochorist
	Dentex dentex	Gonochorist
Maenidae	*Maena* spp.	All protogynous
Labridae	*Coris julis*	Protogynous
	Labrus spp.	Protogynous or gonochorist
Cepolidae	*Cepola* spp	All protogynous
Lethrinidae	*Lethrinus* spp.	All protogynous
Pomacentridae	*Amphiprion* sp.	Protandrous

sexuality. *Sparus* itself includes the important gilt head bream, *Sparus auratus*, of great fish farming potential. It is a protandrous hermaphrodite but not all of the males undergo sex reversal (Zohar *et al.*, 1978), so some genetic control of sexuality remains possible.

Of the other groups in Table 10.1, the Myctophiformes are rare deep sea fish comprising about a dozen families, four of which contain synchronous hermaphrodites. Little is known of the biology of these inaccessible animals, but it seems reasonable to assume that synchronous hermaphroditism is an adaptation to life in the deep sea, and in parallel with the parasitic male tendency in the deep sea angler fishes, serves to guarantee sexual contact in a cold, dark and bland environment.

The swamp eels of the family Synbranchidae are freshwater species widely distributed in tropical regions of the world and contain two species, *Monopterus alba* and *Synbranchus bengalesis* which are protogynous hermaphrodites (Liem, 1968). Finally, a further oddity is the single hermaphrodite species known in the order Cyprinodontiformes, *Rivulus marmoratus*. This remarkable self fertilizing hermaphrodite species has been thoroughly studied by Harrington (1961, 1975). The gonads contain separate ovarian and testicular regions, and although there is no common duct between them the fact of self fertilization is not in doubt; Kallmann and Harrington (1964) have demonstrated that the species exists as a collection of homozygous 'clones', which would be the inevitable fate of a species locked into obligatory inbreeding by self fertilization. Under some circumstances, however, these normally female-like fishes turn into males (Harrington, 1971), so their evolutionary cul-de-sac may be avoided.

Sadovy and Shapiro (1987) review the methods used to diagnose hermaphroditism in fish and emphasize that some criteria are more dependable than others. Differential size frequency distribution for the sexes, for example, is a natural consequence of protogyny but it can also arise in some gonochorists by other routes, e.g. dimorphism as a secondary sexual characteristic. More reliable indications of hermaphroditism are the functional co existence of both types of gonadal tissue or, the most common occurrence, the transitional state in sequential hermaphrodites when one sex becomes transformed into the other. It is this phenomenon of sex reversal which has dominated the study of sequential hermaphroditism in fishes.

The pattern of gonad structure and the sequences of transitional events in sequential hermaphroditic species is very varied and clearly has evolved independently within taxonomic groups and within the different forms of sexuality. Three basic gonad structures are found (Sadovy and Shapiro, 1987) and are illustrated in Fig. 10.1. In the simplest form (Fig. 10.1(a)), separate ovarian and testicular organs are delimited and when sex change occurs, the redundant organ disappears almost completely. In the other two cases the gonadal tissues are not separate but exist either as separate zones within the bisexual gonad (Fig. 10.1(b)), or as islets of the alternate phase spread throughout the current functional gonad (Fig. 10.1(c)). Under both of these systems the process of change involves one tissue expanding and infiltrating

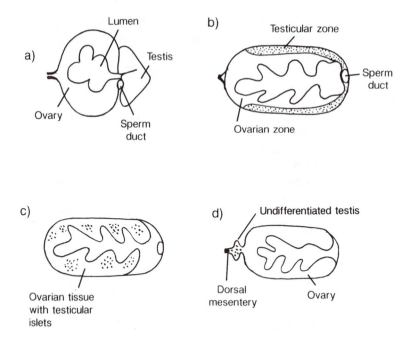

Fig. 10.1 (a-c) The basic features of the gonads of functional hermaphrodite fish (after Sadovy and Shapiro, 1987). (d) A postulated form of the bisexual state in *Betta splendens*.

to replace the other. Residual degenerate material of the redundant gonad phase can be found later in the reversed functional gonad and is diagnostic. In addition, a distinction between protogyny and protandry can be seen in the undelimited gonadal succession in that, in the former, the female lumen and ducts remain to be incorporated into the testicular system whereas in the latter, the male ducts (Wolffian) degenerate to be replaced by the female duct proper (Müllerian). We shall note later that artificial sex reversal from female genotype to male phenotype in trout can be complicated by abnormal or absent male ducts. It is not known whether these phenomena in natural and artificial sex change are related. A further anomaly is the reversion to maleness described by Lowe and Larkin (1975) in *Betta splendens* following surgical removal of ovaries. Both lobes of the ovary and the oviduct were removed but functional males still developed in about 50% of the experimental fish. This implies that testicular material and associated duct primordia are quite distinct from ovarian structures (Fig. 10.1(d)).

The mechanisms underlying the initiation of sex reversal are still not fully understood but some significant advances have been made recently. Thus social behaviour is clearly implicated in sex reversal in some fish. Fishelson (1970) first demonstrated that removal of a dominant male in a population of *Anthias squamipinnis* resulted in the masculinization of one of the remaining females, and subsequent studies have confirmed and expanded this principle (Ross *et al.*, 1990) to include social groups as well as individualists. Several different phylogenetically orientated models have been advanced for the social behaviour component of the process of induction of sex reversal (Ross *et al.*, 1990) but no physiological pathway has been identified. One significant feature of sex reversal is that it can occur at almost any time although it mostly occurs after the reproductive season. Bruslé (1987) followed sex reversal of *Coris julis* throughout a year. These were amongst the earliest of fish recognized as hermaphrodites, and the male and female forms were once thought to be different species. Bruslé found transitional fish with reverting gonads in spring, summer and winter, although the largest number occurred in late summer, after the main reproductive season.

The lack of strict correspondence between sex reversal and the reproductive season, and the fact that social cues can initiate the process in some species, suggest that the endocrine system itself is not part of the sex reversal induction system. This is further supported by experiments involving treatment of fish with sex hormones in which secondary sexual characteristics of the fish could be affected but not the state of the gonad. Chan *et al.* (1972) failed to detect changes in the ovaries of *Monopterus alba* females into which androgen implants were made, and Tang *et al.* (1974) likewise found no effect of injected androgens. A similar result was obtained by Kramer *et al.* (1988) in the protogynous wrasse *Thalassoma bifasciatum* using implants to prolong exposure. Massive degenerative changes occurred in the ovaries but no spermatogenic development was observed. The appearance of blue coloration, a secondary sexual characteristic of the fish, confirmed the masculinizing effect at that level but no primary change was observed in the gonads. In one sense, this corresponds to the effect of androgens on gonochoristic fish which have passed the sexually indeterminate phase. They too become masculine in appearance but no gonadal change occurs.

There is a possibility that functional oogonia exert some block to the progress of spermatogenesis in the wrasse experiments. Kramer *et al.* (1988) reported that oocyte degeneration was marked but did not observe the fate of oogonia. In Chapter 12 we shall see that triploid fish (with three sets of chromosomes) are fully sterile in females in the sense that little or no ovarian tissue develops, nor any gonadal steroids; in the triploid male, however, spermatogenesis does proceed almost normally, although chromosome imbalance generates full sterility here too, and fertilized eggs are not viable. In triploid protogynous hermaphrodites there would be no inhibiting effect of ovarian tissue or oestrogenic steroids, and male inducing factors could therefore be studied in isolation. This looks to be a promising line of research which has so far not been attempted.

Reinboth (1988) concludes that the lack of understanding of the physiology of sex reversal within hermaphrodites may be due to "the fact that ambisexual fishes – most of which are living in the sea – are not very suitable as laboratory animals". This state of affairs should be coming to an end with the expanding use of hermaphrodite species such as *Sparus auratus* for fish farming and the continuing development of rearing methods for the coral reef fishes now so popular with aquarists. These opportunities allied to the potential of chromo-some engineering and gene manipulative methods should open the way for a more effective approach to the endocrinology and genetics of ambisexual states in fish.

One final attempt to identify a switch mechanism for sex reversal which could facilitate the changeover, following behavioural and social cues, with-out implicating the endocrine systems in its usual form is the suggestion by Shapiro (1988) that the H–Y antigen could be involved. The involvement of this phenomenon in sex determination was described in Chapter 6. It initially appeared to be an antigen specific to the heterogametic sex in a wide range of organisms throughout the animal kingdom. Shapiro cites work by Pechan *et al.* (1986) in which H–Y antigen was shown to develop during the sex reversal phase of *Anthias squamipinnis* females after the dominant males had been removed. Shapiro suggests that this might reflect a direct consequence to genetic DNA of the action of the behavioural cues, but it seems more logical that this demonstration of H–Y antigen in hermaphro-dite fish adds credibility to the idea that the antigen is derived and not in itself fundamental (Simpson, 1986).

10.3 ABNORMAL HERMAPHRODITES

These occur as individual fish containing ovarian and testicular material within species which otherwise display the normal gonochoristic sexuality. Abnormal hermaphrodites are found sporadically in a wide taxonomic range of fishes and their frequency in individual taxa owes more to the opportunity for observation than to biological factors. One reported discovery, possibly apocryphal, occurred at the dinner table!

From the genetic standpoint, the most interesting examples of abnormal hermaphroditism are those where the fish can be used for breeding and contribute to our understanding of the genetics of intersexuality and to studies of inbreeding. Self fertilization is common enough in plants and is the most extreme form of inbreeding, but it is not common in animals and almost unknown in vertebrates outside the teleosts.

One of the earliest observations of self fertilization in abnormal hermaph-roditism was first mistaken for an example of 'virgin birth'. Spurway (1953) noticed that some Guppy virgins being stored for future genetic crosses gave birth to broods of offspring. Male contributions were ruled out and the early intimation of virgin birth generated considerable public interest – as it still does today! Further research, however, indicated that the fish contained ovarian and testicular material and the explanation for the observed births was that the fish in question were self fertilizing hermaphrodites (Spurway, 1957). A similar case of abnormal hermaphroditism has been reported in *Fundulus diaphanus* (Porter and Fivizzani, 1983), another cyprinodont com-monly bred in captivity. One individual was found to contain mature sperma-tozoa and ova in the ovarian lumen. Fertilized eggs were not found but the proximity of normally mature eggs and spermatozoa in the fish at the normal time of the year for the species suggests that fertilization was feasible.

Spontaneous hermaphrodites are reported frequently in salmonids and related coregonids and clupeids. O'Farrel and Peirce (1989) discovered a sea trout, *Salmo trutta*, which shed both eggs and spermatozoa. Eggs were fertilized by sperm from another fish and from the hermaphrodite. The crossbred eggs had a 38% survival to the eyed stage just prior to hatching whereas the self fertilized eggs achieved only 7% survival. Complete failure of self fertilization was possible as the authors comment that accidental fertilization in the incubating troughs by contamination from other artificial fertilizations was feasible.

Experimental production of hermaphrodites by accident is a common occurrence in work on sex ratio control in salmonids and tilapias. Details will be given in the next chapter, but part of the process involves masculinization of genotypic females, and if this is not done carefully, sex reversal is incomplete and functional hermaphrodites can arise. The reverse process, feminization of genotypic males, can also lead to the production of hermaphrodites. Jalabert *et al.* (1975) report the production of fry from self fertilizations from such fish.

10.4 FEMALE-ONLY SPECIES

Unisexual or female-only species are found in fish, amphibia and reptiles amongst the vertebrates and they present considerable problems in genetic, ecological and evolutionary terms. That they must be twice as productive as a population comprising equal numbers of males and females, other things being equal, seems undeniable but other things such as fitness and flexibility are not necessarily equal. Unisexual fish occur in several taxonomic groups but first we shall consider a species which is undoubtedly female in form, but is usually described as a self fertilizing hermaphrodite, before dealing with the accepted female-only species.

Rivulus marmoratus

This egg-laying cyprinodont fish (Fig. 10.2) is native to parts of the south eastern United States and Cuba. Its hermaphrodite status was demonstrated by Harrington (1961) who described the gonad as containing mature ova and spermatozoa. These elements were in discrete patches mixed together in the gonad but had no common duct. Despite this, Harrington was firmly convinced that self fertilization occurred. Evidence in support of this was presented by Kallman and Harrington (1964) who tested for the presence of homozygous clones by the use of tissue grafting techniques. Self fertilization is the highest natural form of inbreeding and if a species reproduced by this means over a number of generations it would become homozygous at many if not all loci. Kallman and Harrington collected six fish from their natural habitat and maintained them, and three generations of offspring, in isolation. Subsequent

Fig. 10.2 *Rivulus ocellatus*: a close relative of the self fertilizing cyprinodont *R. marmoratus* (TL about 6 cm).

grafting of fin, heart and spleen material showed that grafts between the generations P_0-F_3 from any one P_0 fish were compatible whilst those across generations from different P_0 fish were not, except in one case. The explanation was that five different homozygous clones were represented in the initial sample.

The tissue graft technique is ideal for assessing the overall relatedness of two or more genomes because it aggregates the effects of all, possibly several, histocompatibility loci. Unfortunately the method does not specify the genotypes in the way that isozyme electrophoresis would. The demonstration of clones is thus not the same as homozygous clones and it remains possible that *R. marmoratus* is not self fertilizing at all but a gynogenetic species of the type to be discussed below (p. 188). One further reason for predicting this is that five different clones were observed in a sample of six fish. It is well known that inbreeding is highly deleterious and if a number of lines, say 30, are started from an outbred base, only one or two are likely to survive 20 or so generations of full sib mating let alone self fertilization. Under the self fertilizing hermaphrodite model, one would predict very few clones unless some mechanism imposed obligate heterozygosity. In *R. marmoratus*, it seems possible that hybridogenesis or some other aspect of chromosomal mechanics could exist.

The *Poecilia* complex

The molly, *Poecilia (Mollienesia) formosa* was the first unisexual fish to be described; Hubbs and Hubbs (1932) believed from the outset that it was of hybrid origin and reproduced by parthenogenesis. In its natural range the species, *P. formosa* is found in the overlapping zone between three other species, *P. latipinna*, the sailfin molly, of coastal distribution (Fig. 10.3), *P. sphenops*, the popular aquarium species, and *P. mexicana* of inland distribution. The hypothesis of hybrid origin arose because *P. formosa* individuals resembled the offspring of the artificial cross between *P. sphenops* and *P. latipinna*. The similarity went no further, however, since the hybrids included males as well as females and reproduced by conventional means. *P. formosa* was found only as females and reproduction required the presence of males of either of the three species of overlapping distribution (Hubbs and Hubbs, 1932; Haskins *et al.*, 1960).

Fig. 10.3 *Poecilia latipinna*: the sailfin molly, implicated in natural hybridization to form all-female species (TL about 6 cm).

Offspring of *P. formosa* were all females and resembled the mother, not the male with which the female mated, even when melanic variants of the males (the black molly) were used. Parthenogenesis was therefore indicated in which the spermatozoa contributed no genetic material to the zygote but were responsible for the activation of the normal egg developmental processes.

As in the case of the hermaphrodite *R. marmoratus*, Kallman (1962a) predicted that broods of *P. formosa* would comprise clones within which tissue grafting would show histocompatibility. This proved correct and in addition Kallman (1962b) showed that clones could also be identified in the wild. Gynogenesis does not necessarily lead to the formation of clones. In the absence of a set of sperm chromosomes, the egg set would normally generate only haploid individuals but somewhere in diploid gynogenesis the egg chromosome number is, in effect, doubled. If this occurs at second meiosis the resulting genomes will be subject to crossing over (p. 131) and the offspring would be variable. In order to develop clones, meiosis must be suppressed at stage I, or alternatively, the haploid complement of the egg pronucleus must be doubled at the first mitotic division. The latter seems improbable, the former is currently accepted for other hybridogenic unisexual fish, and the presence of different clones therefore suggests that the phenomenon by which a hybrid is transformed into an individual which reproduces by ameiotic parthenogenesis is not unique but can happen again and again. Otherwise, i.e. if it happened only once, all of the fish would belong to the same clone and be distinguished only by recurrent mutation, which is slow and random.

A further twist to the tale came with the recognition of occasional paternal contribution to the phenotypes of offspring. Rasch and Balsano (1974) and Balsano *et al.* (1972) were able to demonstrate by the use of electrophoretic techniques that these patroclinous offspring were triploid, having the double chromosome set from the female *P. formosa* and a single set from the *P. mexicana* male. The existence of triploidy was confirmed by chromosome counts (Schultz and Kallman, 1968).

This electrophoretic evidence also confirmed the hybrid origins of the diploid P. *formosa* by showing that its electrophoretic phenotype included the banding patterns of P. *latipinna* and P. *mexicana* – the triploid was confirmed as having one set of *latipinna* genes and two sets of *mexicana* genes. Such elegant confirmation of the older hypotheses (although the hybridization postulated by Hubbs was *sphenops* × *latipinna*) reflects the power of the electrophoretic methods – the picture could now be completed by mitochondrial DNA techniques which would identify maternal complements of hybrids, and evidence on how the triploids breed.

Hybridization, suppression of meiosis, gynogenesis, diploidy and triploidy as observed in the *Poecilia* complex are common and related themes in unisexual species. In another cyprinodontid genus *Poeciliopsis*, the same features exist in more complex form and, in addition, artificial production of the parthenogenetic hybrid has been achieved in the laboratory!

The *Poeciliopsis* complex

The story of this species complex has been reviewed by Schultz (1973a). It was first identified as a unisexual system by Miller and Schultz in 1959 and various studies confirmed the existence within the complex of two normal gonochoristic species, P. *monacha* and P. *lucida*, and three forms of all-female fish intermediate in appearance between the gonochorists. These intermediate forms comprise one diploid, assumed to be the straightforward hybrid, and two triploids whose appearance is biased towards either *lucida* or *monacha* and are therefore assumed to be backcrosses between the straight hybrid and males of one or other of the gonochoristic species. The three forms were therefore named P. *monacha-lucida*, P. *2 monacha-lucida* and P. *monacha-2 lucida*.

The triploids reproduce in a strictly matroclinous mode and can mate with males of P. *lucida* or P. *monacha* or, indeed, a variety of other *Poeciliopsis* species. Whatever males are involved, the offspring are all identical to the mother. The same is true of the cross P. *monacha-lucida* × P. *lucida*, the normal mating, but if the diploid unisexual is mated to males of another species, the offspring resemble the hybrid between P. *monacha* and the male species actually used – the lucida part of the *monacha-lucida* hybrid is eliminated. Any continued backcrossing always leads to hybrids of *monacha* × the last male type. This persistent patroclinous mode of inheritance must involve at each generation the elimination of the paternal chromosomes during meiosis - it is termed **hybridogenesis**.

The explanation of these two forms of sexuality – gynogenesis in triploids and hybridogenesis in the diploid hybrid – was provided by a cytological study of meiosis within the complex (Cimino, 1972a,b). The triploids avoid meiosis by the duplication of chromosomes in the primary oocyte followed by reduction along mitotic rather than meiotic lines to re-establish the original triploid condition. In the diploids, the metaphase plate formed around only one pole to which only the chromosomes of maternal origin were linked – paternal chromosomes were thus eliminated at each generation.

The second major development within the *Poeciliopsis* group was the synthesis of the all-female form (Schultz, 1973b). By crossing *P. monacha* females derived from three different localities with *P. lucida* males also from different sites, F₁ hybrids were produced. All were female and all subsequently behaved as *P. monacha-lucida* in showing hybridogenic reproduction.

As Schultz (1973a) reflects, "The laboratory synthesis ... opens the way for a vast area of enquiry that is virtually unexplored there is no reason to believe that they {unisexual forms} are not still being generated in nature." Likewise there is no reason to believe that properly planned hybridization of species in aquaria conducted by amateur or professional breeders might not also be successful in these endeavours.

The silver carp

Carassius auratus, the forerunner of the goldfish, has two subspecies widely distributed in Eastern Europe, mainland Asia and Japan which are unisexual. *C. auratus gibelio* (Fig. 10.4) (sometimes called the silver carp) is the mainland subspecies and coexists with normal gonochoristic species over much of its range (Cherfas, 1966). It is not known to be of hybrid origin and some authors state that it cannot easily be distinguished from gonochorists of the same subspecies (Vasil'yeva, 1990). In some parts of the former Soviet Union it does appear to differ from the bisexual species in having a pronounced first ray on the dorsal fin, which is indicative of a common carp, *Cyprinus carpio*, involvement. The all-female forms are triploid and gametic formation seems to follow the plan of the *Poeciliopsis* triploids, with duplication to a hexaploid condition in place of first meiosis followed by a simple mitotic-type reduction division to re-establish triploidy. The cytological basis for this was described by Yu (1982) in artificial crosses between silver carp and common carp. There was no polar body corresponding to meiosis I but there was a tripolar spindle and a polar body extrusion after fertilization. A similar cross was reported by Cherfas *et al.* (1985) which also generated sterile male fish. Such anomalies in male

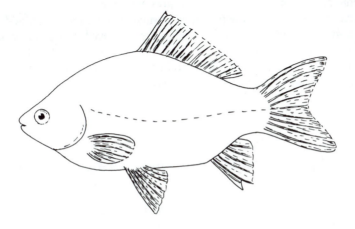

Fig. 10.4 *Carassius auratus gibelio*: the silver carp of all-female form (TL about 16cm)

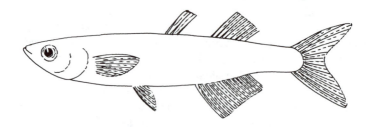

Fig. 10.5 *Menidia*: a marine genus, the silversides, amongst which is found an all-female species (TL about 15 cm).

sexuality in *Cyprinus carpio* also arise in induced gynogenesis (Chapter 12). The Japanese all-female form is *C. auratus langsdorfi* (Kobayashi, 1971) and both it and the mainland form exist as populations of clones (Nakanishi, 1987).

Recent additions to the all-female fish list

The silverside fishes of the genus *Menidia* (Fig. 10.5) are found in brackish waters off the Texas coast. All-female forms called *Menidia clarkhubbsi* coexist with two gonochoristic forms, *M. beryllina* and *M. peninsulae*, but initially (Echelle and Mosier, 1981) were thought not to be of hybrid origin, at least between these two species. Echelle *et al.* (1988) now define the silverside populations as comprising the two gonochorists, their hybrids which are bisexual and the all-female form which possibly represents a hybrid between *M. beryllina* and an unknown ancestor. The unisexual individuals were all diploid but the gonochorists included triploids with two sets of genes from one or other parent or straightforward diploids. The situation is not quite the same, therefore, as in the *Poeciliopsis* story but gynogenetic all-female hybrids with fixed heterozygosity and clonal population structure were features once again.

A further group of fishes within the Cyprinidae also contains an all-female species in which hybridization and polyploidy are involved. The minnow species *Phoxinus eos* (Fig. 10.6) and *P. neogaeus* are common in eastern North America and Canada and hybridize readily in nature. The hybrids are female and constitute a unisexual species (Dawley *et al.*, 1987). Tissue grafting showed the existence

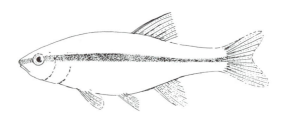

Fig. 10.6 *Phoxinus*: the diminutive minnows; *P. eos* an all-female form (TL about 6 cm).

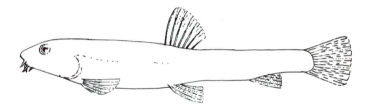

Fig. 10.7 *Cobitis*: the loaches also contain an all-female species, *C. taenia* (TL about 12 cm).

of histocompatible clones, and diploid and triploid individuals could be recognized electrophoretically and identified as containing two doses of one set of genes and one of the other, in both combinations, as within the *Poeciliopsis* situation. An additional feature of this complex is that individuals can be mosaics, with some cells containing the diploid number of chromosomes, others the triploid number. This suggest that chromosome number adjustment as seen in hybridogenic meiosis may also be possible in mitotic divisions.

A further family, the Cobitidae or loaches, contains an all-female species (Vasiliev and Vasilieva, 1982). This case is less well documented than the others but seemingly involves *Cobitis taenia* (Fig. 10.7) and another species inhabiting the Moscow River. The all-female forms are either triploid or tetraploid or even pentaploid.

Finally, again within the Cyprinidae, the bleak species *Rutilus alburnoides* is reported to contain a diploid-triploid all-female group (Collares-Pereira, 1985).

Conclusion

It is obvious that a wide range of possibilities exists for the creation, either naturally or artificially, of unisexual species by interspecific hybridization. The meiotic mechanisms which permit the stable inheritance of fixed heterozygosity have been described, but the initiation of these events is not understood but could be of primary importance in fish breeding.

An exciting but unconfirmed development is the finding that sex reversal of gynogenetic females to males produces fertile fish which produce diploid spermatozoa (Gomel'skii *et al.* 1985). The all-female stocks of *Carassius auratus gibelio* were apparently derived from crossing the bisexual form of this fish with *Cyprinus carpio*. The hybrids were diploid and reportedly produced diploid eggs by replication of chromosomes within the primary oocyte followed by a mitotic division. Thus it would appear that whatever suppresses meiotic reduction during oogenesis must also operate in spermatogenesis. Following from this, the use of sex reversal techniques to generate males in the various all-female species is an obvious line of further research. Attempts to produce tetraploid male fish so as to generate diploid spermatozoa for the production of sterile triploids is a very important goal in applied fish breeding (p. 213). This alternative method offers another promising way forward.

Chapter eleven

Control of sex ratio

11.1 INTRODUCTION

The genetics of sex determination was discussed in Chapter 8 and some aspects of atypical sexuality in Chapter 10. This chapter covers the practical use of measures to control sex ratio in fish breeding. This represent perhaps the most successful application of genetics to fish cultivation. It has transformed some aspects of trout and salmon farming and is probably applicable to the breeding of all fish except perhaps the hermaphrodite species. Even with these, the hormonal techniques which must accompany the genetic procedures of sex ratio control are of considerable practical value, and there must be significant potential overall for the development of all-female hybrid forms which will breed true.

The reason for wanting to control the proportion of males and females in fish breeding is that the sexes in fish are often different from one another in a range of characteristics which have economic relevance. Thus the sexes often display different growth rates. For example, in many species the female grows quicker than the male after sexual maturity has set in. The marine flatfish present good examples of this and in most species the male is very much smaller than the female. In many other families such as the pike (Esocidae) and the eels (Anguillidae) there is a sharp distinction between male and female size, whereas in many species of the Cyprinidae the males are generally smaller than the females but sizes overlap. In other species, particularly where male parental care or aggression are features of spawning, the male is the bigger of the sexes. This is true of many cichlids such as the tilapias and it also seems true for some catfish of the family Ictaluridae (Simco *et al.*, 1989).

The main reason for selecting one sex for cultivation rather than the other lies in the act of reproduction itself, or the secondary consequences of it. In salmon and trout, growth of the males ceases when secondary sexual characteristics appear and the fish develop territoriality and aggression, which leads to loss of condition, poor overall appearance (Fig. 11.1) and unsatisfactory flesh texture and taste. Dark salmon or trout are of little value as food and are despised by sport fishermen. Even more than this, male salmonids may become sexually mature before they are 1 year old. In salmon this is

Fig. 11.1 A sexually mature male rainbow trout, *Oncorhynchus mykiss* (TL about 25 mm).

referred to as precocious sexual maturity, occurs during the freshwater phase of life, and results in a fish of no value for growth in the sea. In rainbow trout such early maturity often occurs where fish are being fed heavily for good early growth. It too represents an economic loss which can wipe out the gain from the good early growth rate of the rest of the fish. The principal motivation for producing female-only trout, however, reflects the use they are put to in sport fisheries. Anglers find the sexually mature males to be pests rather than good quarry, and since this state of maturity can last for half a year in male rainbow trout in put-and-take fisheries their presence there is distinctly undesirable.

The reasons for wanting male-only tilapia have already been stated (p. 145): the females breed prolifically and overpopulation and consequent stunting occur to the detriment of the fishery. By using only males, the tilapia breeder not only avoids unwanted natural reproduction but also ends up with the bigger of the two sexes, which is also an advantage where basically small fish, such as the tilapias, are being reared for consumption.

The scientific basis for the genetic and hormonal control of sex determination in fish was largely defined by the pioneering work of Yamamoto in the 1950s (reviewed 1969b), but its application to fish farming (Purdom, 1977, 1978) generated a much wider interest and it is now almost a standard procedure in advanced fish cultivation and research.

11.2 METHODS FOR SEX RATIO CONTROL

The control of sex ratio by the development of appropriate hybrids was described in Chapter 9. It related entirely to tilapia culture and failed to achieve basic commercial utility because it was not absolute and therefore did not limit natural reproduction. Another method of sex control which could be absolute is the treatment of fish with sex steroids during the sexually indeterminate early phase of life. This too has not been of much commercial value due to several factors: first, there is the need to treat all fish every generation and often under veterinary supervision; secondly, treatment can lead to significant fry mortality; but most of all the use of hormones leads, in the food trade, to consumer resistance. Sophisticated eaters are increasingly averse to foods which are produced using hormone supplements. This leaves the genetic method of sex ratio control, albeit with support from hormonal treatments, as the most satisfactory procedure.

Breeding tactics were derived from the original demonstration by Yamamoto that sex was determined genetically, possibly via sex chromosomes, and that the control was mediated in the early days of feeding after hatching, during which time genotypic sex could be reversed by the application of the appropriate sex hormone. Thus potential males could be feminized by oestrogens and potential females could be directed towards male development by androgens.

Manipulation of XX/XY systems

The production of all-female systems of breeding from species in which the normal mode of genetic sex determination of the XX female, XY male type is shown in Fig. 11.2. Starting with normal fry at the sexually indeterminate stage, the procedure first requires the masculinization of the fish with androgens. The resultant males will comprise roughly equal numbers of XX and XY individuals, and these are progeny tested by crossing with normal females: the P_0 fish identified as XY males, by the fact that their offspring include males, are discarded. The others (XX) are retained and further crossed with XX females to generate in the F_1 an all-female stock from which the Y chromosome has been eliminated. To continue into F_2 and beyond, it is necessary to masculinize a proportion of the fish each generation. In practice, in salmonids, the sex reversed males of XX constitution were not quite normal in that the sperm ducts were incomplete. This had two consequences, one good, one bad: the good part was that prior to progeny testing, the operator had a good guide to the likely outcome of the test crosses; secondly, however, the more significant bad part was that the male had to be killed to obtain milt, and sufficient F_1 offspring therefore had to be masculinized to continue the process into F_2 and beyond. In practice, the selection at F_1 was of broods of masculinized offspring from XX males whilst those from XY males were discarded.

This simple procedure for establishing a female-only stock for fish of the XX female, XY male type has been repeated many times. The only difficulties

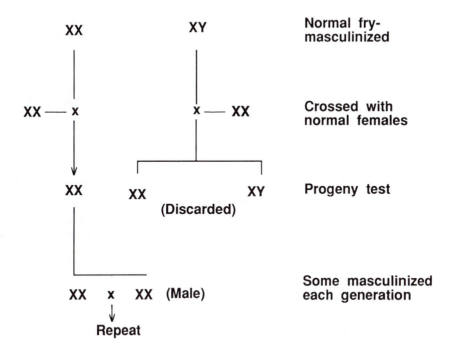

XX XY	Normal fry-masculinized
XX ── x x ── XX	Crossed with normal females
XX XX XY	Progeny test
(Discarded)	
XX x XX (Male)	Some masculinized each generation

Repeat

Fig. 11.2 Schedule for all-female production in an XX female, XY male sex control system.

in salmonids are the progeny testing which is only necessary in the first generation anyway, and the ductlessness of males which means they must be killed to obtain milt – this will be reconsidered later.

Production of all-male systems is not as simple as that for all-females, requiring one more generation to set up the system. The basic procedure is shown in Fig. 11.3 and starts in much the same way as the all-female system. Normal fry are feminized with oestrogens and after crossing with normal males, the females of genotype XY are identified as those which produce a 1:3 sex ratio in favour of males. From here on the techniques are more demanding than those shown in Fig. 11.2. The simplest procedure is to feminize a proportion of all F_1 offspring, eventually discarding those from which the test progeny indicates a 1:1 sex ratio. Of the other group, the non-feminized batch will include two sorts of male, XY and YY, whilst the feminized group will have females of all three genotypes, XX, XY and YY. By crossing in all combinations, the desired cross of YY × YY will produce all males of which a proportion must be feminized to continue the stock. The all-male production line may then be run as the mirror image of the all-female system with YY males for production purposes and feminized YY individuals to continue the line. An alternative would be to use a cross between YY males and normal (XX) females to produce XY males – i.e. to maintain two parental lines, one

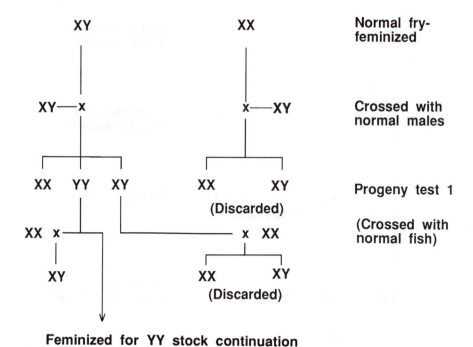

Fig. 11.3 Schedule for the creation of YY stock for an all-male production system where sex control is of the XX female, XY male type.

normal and the other of YY stock with F_1 progeny for the output side. The advantage of this is that the XY hybrid would probably be fitter than YY males but, more importantly, the XY F_1 hybrid would have no breeding value in terms of sex ratio control and the economic benefit of constructing the all-male system would be protected.

It is also possible to generate YY individuals directly by androgenesis (see Chapter 12) or by a combination of sex reversal and gynogenesis. In diploid androgenesis, the objective is to remove or destroy the female egg chromosomes and after fertilization double up the spermatozoan chromosomes at first mitosis. From a Y-bearing sperm cell a YY individual would be produced directly. Combining sex reversal and gynogenesis for the same purpose entails sex reversal of a genotypic male (XY) followed by diploid gynogenesis to produce YY homozygotes. Both gynogenesis and androgenesis will be covered in greater detail in the next chapter. Application of these novel techniques has not yet been achieved in commercial practice but the production of YY males in *Oreochromis mossambicus* by sex reversal and meiotic diploid gynogenesis has been reported (Varadaraj and Pandian, 1989).

Male-only systems do not yet seem to be in commercial use but they offer advantages in some fisheries, e.g. tilapias generally, channel catfish farms and brown trout sport fisheries where specimen fish tend to be male – in sharp contrast to the situation already described for rainbow trout.

Manipulation of WZ/ZZ systems

In theory exactly the same set of procedures, but in mirror image, could be used for monosex production in fish where sex determination is of the WZ female type. Clearly the homozygous all-male procedures are simpler than the all-female ones in this case, and species for which the techniques should be explored include flatfish such as turbot, *Scophthalmus maximus*, and those tilapia species showing female heterogamety such as *Oreochromis macrochir*. In fact, it seems logical, if all-male production is of vital importance, to choose a species with female heterogamety so as to be able easily to construct a genetic breeding system for all-males. The repeated use of direct masculinization is then avoided, with gains on economic, aesthetic and genetic grounds.

Hormone-induced sex reversal

The widely varying procedures for masculinizing or feminizing fish by the use of sex hormones have been reviewed by Yamazaki (1983) and Hunter and Donaldson (1983).

Three basic ways of administering sex hormones for the purpose of sex reversal are by injection, by immersion and by feeding. Of these the last is the most practical approach, and simply requires that appropriate levels of hormone be incorporated into normal diets, which are fed to the fish during the relevant period and at an appropriate rate. The variables are therefore the nature of the hormone, its concentration in the diet, and feeding rate and duration.

Yamamoto (1969b) lists a range of oestrogens and androgens used in the seminal Japanese work. Of the androgens, 19-nor-ethynylestotestosterone was the most potent, but not far behind was the artificial analogue methyltestosterone which is cheap and readily available: this is the usual androgen used in practical masculinization. There was less variation in potency of oestrogens and the commonly available oestradiol-17β and ethynylestradiol were both highly effective.

Hormone is usually incorporated into food as a solution in ethyl alcohol which is subsequently evaporated off. Dietary concentrations were defined by Yamamoto in terms of doses required to achieve 50% sex reversal but the more common requirement in practical breeding is the minimum dose to achieve complete sex reversal. By trial and error a range of dietary concentrations and feed times has been developed for a number of fish species (Tables 11.1 and 11.2). There is an increasing tendency towards lower concentrations of hormone. The normal level of 3 mg per kg of food was used most often for sex reversal in rainbow trout and generated virtually 100% males from genotypic females, but even lower doses are used now. Much of the work with

Table 11.1 Masculinization of salmonids by feeding with methyltestosterone

Species	Concentration (mg kg^{-1})	Duration (weeks)	Effect (%)	Reference
Oncorhynchus mykiss	15–60[*]	20	60%	Jalabert *et al.* (1975)
	3	13	80%	Johnstone *et al.* (1978)
	1–10	8	70%	Okada *et al.* (1979)
	1[*]	6	100%	Bye and Lincoln. (1986)
O. kisutch	20[*]	10	sterility	Goetz *et al.* (1979)
	10	12	sterility	Hunter and Donaldson (1983)
	1	9	60%	
O. tshawytscha	3–9	3–9	90%	Hunter *et al.* (1983)
Salmo salar	3	13	100%	Johnstone *et al.* (1978)

[*] Including immersion treatments. See text for explanation.

cichlids was aimed at direct masculinization rather than at breeding methods involving genetics, even though the early work by Clemens and Inslee (1968) had shown the potential for the genetic approach. Thus the massive overdose was acceptable because fish were not expected to breed and sterility was no problem. With the additional requirement that the males must be fully fertile, the lower doses became more practical (Pandian and Varadaraj, 1987). Even greater reductions in hormone concentration are now used in trout and salmon work. The reason for this is that the earlier treatment, involving diets containing 3 ppm of hormone or more, produced males lacking sperm ducts (Fig. 11.4), and this was inconvenient because the males had to be killed for use. This not only meant that male fish had to be raised anew each year, but also that males might be killed before the testis was fully ripe. In addition, because of the lack of sperm duct glands, the seminal fluid was inadequate and diluent, such as those used in sperm cryopreservation (p. 23) had to be

Table 11.2 Masculinization of cichlids with dietary methyltestosterone

Species	Concentration (mg kg^{-1})	Duration (weeks)	Effect (%)	Reference
O. mossambicus	30	10	100%	Clemens and Inslee (1968)
	50	3	100%	Nakamura (1975)
O. aureus	15–60	3	80-100%	Guerrero (1975)
	60–240[x]	3	99%	McGeachin *et al.* (1987)
O. niloticus	40	9	100%	Jalabert *et al.* (1974)
	5	6	100%	Owusu-Frimpong and Nijjhar (1981)
	30[x]	8	100%	Tayamen and Shelton (1978)
O. zillii	50	7	100%	Woiwode (1977)
O. macrochir	40	8	sterile	Jalabert *et al.* (1974)
O. mossambicus	5	2	100%	Pandian and Varadaraj (1987)

[x]ethyltestosterone

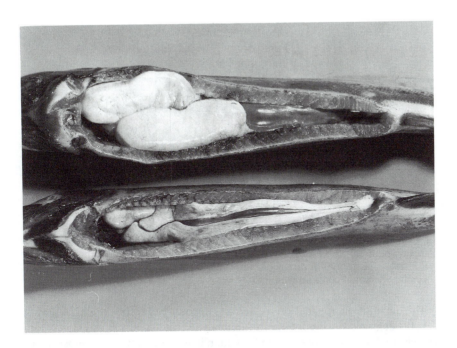

Fig. 11.4 Normal rainbow trout male with sperm ducts and ductless sex-reversed male (TL about 25 cm).

employed. Thus there was good reason to test a lower dose regime in order to produce more normal males.

Bye and Lincoln (1986) used methyltestosterone at 1 mg kg^{-1} in the diet and treated for 450 degree days (i.e. temperature in °C × days) to achieve 100% masculinization of rainbow trout with a low level of normal males from which milt could be stripped. An extension of this work at 0.5 mg kg^{-1} over a range of times was even more successful (Fig. 11.5) and produced almost 60% normal males. Confirmation of this is provided by Cousin-Gerber *et al.* (1989), who produced over 80% fully functional male rainbow trout using 0.5 mg kg^{-1} of methyltestosterone in the diet, fed for 60 days from first feeding. The final problem with sex ratio control in rainbow trout seems to have been solved and although reversal is less than 100%, the loss is not significant and the absence of genotypic males avoids any confusion between reversed and natural males. At the lower range of doses used in tilapias, there appear to be no problems with male fertility.

Methods other than dietary administration include treatment by immersion and by injection. Much of the immersion work has been done in conjunction with oral administration and is probably unnecessary in practical use. Treatment by injection in grass carp was successful, however (Shelton, 1986), under circumstances where feeding was ineffective. The hormone was put in lengths of silastic tubing and implanted into fish of length 65 to 165 mm.

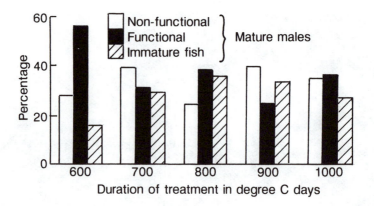

Fig. 11.5 Histograms showing the frequency of (black columns) fully functional males, (open columns) non-functional ductless males, and (hatched columns) immature fish following masculinization of rainbow trout fry with methyltestosterone (R.F. Lincoln, personal communication).

The duration of application of the diets appears not to be critical. A period of 1000 degree days is effective in salmon and trout, and various times from 3 weeks to 2 months work in cichlids which live at much higher temperatures than salmonids. The phrase 'to treat during the sexually indeterminate phase' is a bit of a circular argument since sex reversal is often used to define that phase. In general it appears that the most important aspect of hormone feeding is to start as early as possible in the fry's first-feeding period.

Administration of oestrogens follows the pattern laid out for androgens, but reversal to femaleness is less reliably produced than reversal to maleness and higher doses are needed. A selection of successful examples in salmonids and cichlid fish species is given in Tables 11.3 and 11.4. One notable difference between the techniques for feminization and masculinization is that the immersion of fry in solutions of oestrogens is often used as an adjunct to dietary administration prior to the onset of feeding, and in some cases, e.g. Nakamura

Table 11.3 Feminization in salmonids by the use of oestradiol-17β

Species	Concentration (mg kg^{-1})	Duration (weeks)	Effect (%)	Reference
Oncorhynchus mykiss	50-100	8	Low	Okada et al. (1979)
Salmo trutta	20	9	100%	Johnstone et al. (1978)
Salmo trutta	20[+]	6	100%	Johnstone et al. (1978)
O. kisutch	10[+]	10	90%	Goetz et al. (1979)
O. kisutch	5[+]	13	90%	Hunter et al. (1982)
O. masou	–[+]	3	100%	Nakamura (1981)

+ including immersion treatments

Table 11.4 Feminization of cichlids using hormone diets

Species	Hormone	Treatment	Effect	Reference
Oreochromis aureus	oestradiol	25–200 mg/kg	60%	Hopkins *et al.*(1979)
	ethynyloestradiol diethylstilboestrol	5–8 weeks		
O. mossambicus	ethynyloestradiol	50mg kg^{-1} 3 weeks	100%	Nakamura and Takahashi (1973)
	diethylstilboestrol	100 mg kg^{-1} 2 weeks	100%	Varadaraj (1989)
O. niloticus	oestrone	100–200 mg/kg 9–11 weeks	70%	Tayamen and Shelton (1978)

(1981), is effective on its own. Concentrations used in the immersion techniques vary from 0.5 µg l^{-1} to 400 mg l^{-1} but the higher levels are for acute treatments, the lower ones below 20 µg l^{-1} for more or less continuous exposure of the alevins.

One theory on the relative ineffectiveness of oestrogens is that breakdown is rapid in the fish tissues, necessitating continuous dosing which normal daytime feeds would not supply. One test of this might be to assess more fully the use of the synthetic steroid 17α-ethynyloestradiol, which is not glucuronated quite as rapidly as normal analogues and is therefore relatively stable within the body.

Most of the work on sex reversal in commercial fish has been done on salmonids and cichlids. Little has been done with cyprinids or catfish, presumably because sex ratio is not a significant feature of the culture of either. Limited work with grass carp already referred to (Shelton, 1986) suggests that they may be responsive to implanted hormones, and the feeding of methyltestosterone to common carp was effective (Nagy *et al.*, 1981; Komen *et al.*, 1989) when a high concentration of hormone, 50–100 mg kg^{-1}, was applied over a period of 4 or 5 weeks. Carp did not respond to oestrogens.

Sex reversal by administration of hormones during the early feeding phases seems to be achievable in most fish species, but the treatment parameters need to be established empirically and a wide range of effective treatments seems likely to be found.

11.3 COMMERCIAL APPLICATION

A very large amount of research is done world-wide in the name of fish cultivation, but rarely is much effort put into translating the science into commercial practice. A deliberate policy decision was taken by the Directorate of Fisheries Research of the UK Ministry of Agriculture Fisheries and Food in the early 1970s to transfer the applied research it undertook in trout farming to the then fast-growing industry. The basic programme comprised stock choice for year-round spawning, the introduction of all-female methods and the development of chromosome engineering techniques for the production of sterile fish (Chapter 12). The success of this programme, particularly in relation to monosex culture, has been reported by Bye and Lincoln (1986) and a brief review is given here.

The principal objective in this work was to encourage active collaboration between farmers, particularly those with significant broodstock and hatchery facilities, and the scientists working in the Directorate. Once the laboratory trials had established the nature of the breeding tactics and hormone dietary procedures required for controlling sex ratio, the work was transferred to commercial sites and carried out there by government scientists. Starting in 1979, the first masculinizations by farm staff were supervised and monitored by scientists who followed the sex-reversed fish through from first feeding up to use as two-year-olds for breeding. The technique for removal of milt and use of diluents was demonstrated at each farm. At the same time, other farms were provided with ready-to-use sex-reversed male fish and the concept of specialist providers of these was promoted. From the all-female ova generated by these first steps, farmers were encouraged to develop sex-reversed fish for their next generation.

The first commercially available all-female trout came on the market in 1980 and interest within the UK was immediate. A widening range of fish farmers participated in the government/industry collaboration and, at the same time, government scientists visited other farms, colleges of agriculture and other education establishments to describe the new methodology and its advantages. This was backed up by a free consultancy service and by the presentation of explanatory papers in the commercial press (Bye and Lincoln, 1979, 1981; Bye, 1982).

There were difficulties. Many farmers required repeated assurance that the females were perfectly normal and that the only significance of the methodology was the absence of males in the production line. Other concerns were that parental treatment with hormones would be environmentally dangerous and that customer resistance to hormone-mediated food production would arise. Both worries were shown to be without foundation. Technical problems were the lack of ducts in the sex-reversed males, already dealt with, and the general availability of male broodstock in a system where natural males were rapidly being supplanted by sex-reversed individuals. The latter difficulty is still occasionally encountered, but the solution is organizational and the need is for the farmers to look ahead.

The final stage of government involvement was to conduct a postal survey to assess the extent to which the techniques for all-female production were being employed. A summary of this survey conducted in 1986 is shown in Table 11.5. Over 70% of responding farms in England and Wales used the

Table 11.5 1985/6 Survey of all-female production in trout farms in England and Wales

Farms			Egg Production		
Style	No.	%	Type	No. $\times 10^6$	%
Traditional	19	28	Mixed sex	10.4	17
Female-only	49	72	Female only	49.5	83

all-female technique, and this included the larger operators because over 80% of eggs produced in these farms were of the all-female type.

It may be argued that the government effort to promote the use of these novel techniques cannot be assessed in the absence of a control. However, the scientific basis for the developments was almost 20 years old and had not, of itself, generated any commercial interest. Fish farming generally comprises small businesses and as such they are unlikely to generate the research and development necessary for such developments, nor to promulgate them if they have made significant discoveries!

Chapter twelve

Chromosome engineering

12.1 INTRODUCTION

Chromosome engineering in fish by the manipulation of chromosome sets developed out of work done in the early years of the 20th Century on the effects of ionizing radiation on frog spermatozoa. The seminal discovery was made by Hertwig (1911), who fertilized frog eggs with irradiated spermatozoa in order to examine the relationship between the dose of radiation applied to spermatozoa and the rate of survival of the embryos. Not content when a 100% lethal dose level was recorded, Hertwig went much further and examined the effect of extremely high doses of radiation and discovered that beyond a certain level (1000 Gy) there was a return to a low level of survival of embryos. This 'paradoxical' effect was rightly interpreted as arising from the complete destruction of the genetic content of the spermatozoan, which permitted the egg nucleus to undergo maturation and mitotic divisions unhampered by the bits of broken chromosome left behind after the lower levels of radiation dose. Thus the eggs began development effectively without a male genetic input, and hence the process has been called **gynogenesis** and is a special form of parthenogenesis.

Additional work by several authors confirmed first that the surviving embryos were mostly haploid, having developed from the single set of chromosomes in the female pronucleus; secondly, that a small proportion of these embryos were normal in appearance and were diploid; and thirdly, that this low frequency of diploids could be increased dramatically by the application, shortly after fertilization of the egg, of various treatments such as cold shock.

In addition to gynogenesis, a complementary process, **androgenesis**, was also established by this early amphibian work. In androgenesis, the egg chromosomes were inactivated by irradiation, and subsequent embryonic development, following normal fertilization, involved only the chromosome set contributed by the spermatozoan.

This extraordinary work in amphibia was reviewed by Fankhauser (1945) and Beatty (1964). It remained only of academic interest with relatively little further development until taken up by fish geneticists and given a new impetus in connection with fish farming. But first, it received further attention from the Soviet school of radiobiology (Romashov *et al.*, 1961), who extended the basic findings of the early amphibian work to include the manifestations of most of the phenomena in fish. The genetic implications, particularly in connection with fish farming, were explored by Golovinskaya (1968) in the Soviet Union and by Purdom (1969) in the West. They were subsequently expanded to include induced polyploidy and taken up by fish geneticists world-wide in pursuit of genetic technology for use in fish cultivation in all its forms. Reviews of these developments have been made by Cherfas (1981), Purdom (1983) and Thorgaard (1983b).

The critical feature of gynogenesis for its genetic consequences was the manner in which diploidy was restored to the otherwise haploid status of the egg. The earlier work with amphibia had not resolved this problem but it was clarified quickly in fish studies. Three ways in which diploid status could be reached in the gynogenetic zygote were by suppression of meiosis I, by failure to eject the second polar body in meiosis II or by failure of telophase and anaphase after the first mitotic division of the zygote nucleus. As described in Chapter 10, natural gynogenesis in all-female forms of fish does entail suppression of meiosis I and leads to obligate heterozygosity and clone formation. This seemed unlikely to be the case for induced gynogenesis because this meiotic phase is completed long before the events which lead to induced diploid gynogenesis following egg fertilization with genetically inert spermatozoa. Both of the other methods had been claimed in the amphibian studies (Beatty, 1964), but on genetic evidence, Golovinskaya (1968) demonstrated that retention of the second polar body was the likely origin of diploidy in fish gynogenesis, and Purdom (1969) reached the same conclusion by observing no cleavage delay in flatfish eggs destined to become gynogenetic diploids. In a final demonstration of this in the loach, *Misgurnus anguillicaudatus*, Oshiro (1987) observed cytologically the

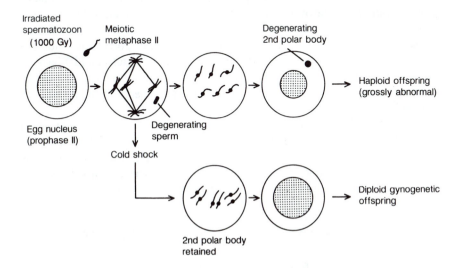

Fig. 12.1 Diploid gynogenesis by the meiotic pathway and cold shock.

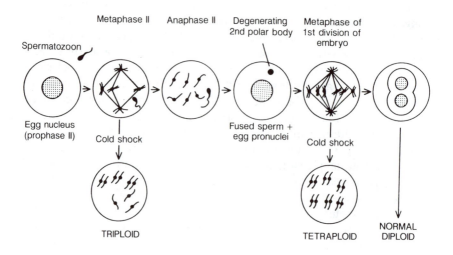

Fig. 12.2 Induced polyploidy by cold shock during meiosis (triploidy) or mitosis (tetraploidy).

relevant events in the fertilized egg. Prior to cold shock, given a few minutes after fertilization, meiosis II had progressed to anaphase II. During the cold shock this phase remained static except that the spindle fibres disappeared, and after the cold shock the two sets of chromosomes reunited to form a single nucleus.

It is now generally accepted that gynogenetic diploids produced by cold shock or other means applied soon after fertilization arise by retention of the second polar body chromosome set. Nevertheless, diploidization by union of the first mitotic products can occur. The overall processes in gynogenesis are indicated in Fig. 12.1.

By employing the same phenomena, but using genetically active spermatozoa, chromosome sets can be manipulated to produce polyploids. Thus fish with three sets of chromosomes (triploids) or four sets (tetraploids) became possible and were immediately of appeal in fish farming as a means of developing sterile fish (Purdom, 1972b). The basic mechanisms in induced polyploidy are illustrated in Fig.12.2.

The critical features of the techniques for chromosome set manipulation are (a) the means or source of genetically inert spermatozoa, (b) the nature of the post fertilization shock treatment, (c) the timing and duration of this treatment, and of course (d) the species of fish to be worked with. These and some applications are dealt with in the succeeding sections.

12.2 MEIOTIC DIPLOID GYNOGENESIS

This is the most straightforward of the chromosome manipulation procedures in that the distinction between success and failure of diploidization is obvious. The appearance of haploid embryos is highly characteristic in all of the lower

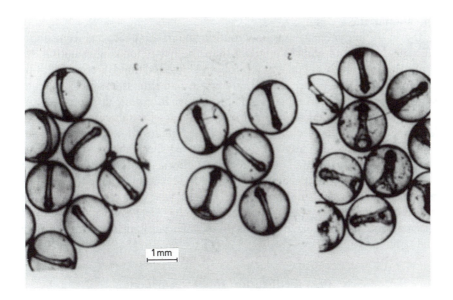

Fig. 12.3 Haploid (a), gynogenetic diploid (b) and normal diploid embryos (c) of the plaice (*Pleuronectes platessa*).

vertebrates and comprises severe malformation of the head with foreshortening of the body and more especially, the tail. The difference between haploid and diploid fish embryos which is clear from even a casual glance (Fig. 12.3), makes the system ideal for experimental purposes.

Inactivation of spermatozoa

The use of ionizing radiation to neutralize the genetic material of the spermatozoan has drawbacks. The required doses are very high and to apply them in a short period means that the source of γ-rays or X-rays must be very powerful and therefore somewhat hazardous and certainly not readily available. An important advance was made when the inactivating effect of ultraviolet (UV) radiation was established. Stanley (1976) first used the method for inactivating grass carp spermatozoa and Chourrout (1982) demonstrated its efficacy also for spermatozoa from salmonids. A quantitative study by Taniguchi *et al.* (1986) using the ayu, *Plecoglossus altivalis*, suggests that in thin layers of milt an absorbed dose of 6000 erg mm^{-2} is the minimum requirement but many workers use levels around 9000 ergs mm^{-2}. Provided that the wavelength is within the bacteriostatic range (i.e. will inhibit the growth of bacteria, but not kill them), an empirical approach to achieve the right dose is to assess that exposure time which kills all spermatozoa and use something a little less than this, which leaves the sperm cells motile. The only other requirements when using ultraviolet irradiation equipment are to keep the exposed milt cool and, of course, observe the safety regulations for operation of the UV lamp.

Use of UV in gynogenesis is now routine and in some respects seems more efficient in terms of embryo survival than the use of ionizing radiation. No firm data exist on this issue, but it has been suggested that UV cross-links the chromosomal DNA rather than shattering it, and thus is less likely to leave small fragments of DNA which could be reactive in mitotic divisions.

An even more trouble-free source of spermatozoa for inducing gynogenesis is to use milt from a species unrelated to the females that provide the eggs. This is frequently done anyway when irradiated spermatozoa are employed so as to use the absence of paternal genetic characters to positively exclude paternal contribution. It is also a feature of the breeding of all-female fish by natural gynogenesis (Yu, 1982), but the only clear case reported so far for induced gynogenesis was the use of halibut, *Hippoglossus hippoglossus*, spermatozoa to activate gynogenesis in eggs of other flatfish species (Purdom and Lincoln, 1974).

Post fertilization shock treatment

Cold shock was the traditional way of achieving diploid gynogenesis in amphibia and served well for the early fish studies with carp, loach, sturgeon and marine flatfish. It is easy to apply precisely by using a mixture of ice and water as a water bath. Unfortunately, it was not appropriate for that important group of fishes, the salmonids (Lincoln *et al.*, 1974), and the early work had to rely on spontaneous diploidization, which seems common in gynogenesis but only at a level of 1% or less of embryos. The situation was radically changed by Chourrout's demonstration (1980) that heat shocks were effective in salmonids. This was quickly confirmed by Thorgaard *et al.* (1982) and Lincoln and Scott (1983) and is now used routinely world-wide for diploid gynogenesis and induced polyploidy. Optimum temperatures for different salmonid species are around 28°C and it seems probable that other families of fish will have different temperature optima, in this context, which will need to be established empirically.

A third type of treatment which had been used in the earlier amphibian work and which was tested in fish from the outset was hydrostatic pressure. Early attempts to develop this method were not successful but Lou and Purdom (1984a) reported the production of diploid gynogenetic rainbow trout using a pressure of 9000 lb inch^{-2} and such effects have been confirmed in the zebrafish, *Brachydanio rerio*, by Streisinger *et al.* (1981) and the medaka by Naruse *et al.* (1985). It was believed that pressure might have fewer side effects than heat or cold shocks but this does not appear to have been borne out by work to date. The technique certainly suffers from severe technical requirements in the construction of the equipment.

Cytostatic chemicals such as cytochalasin-B are theoretically suitable for chromosome set manipulation. Limited study suggests that such chemical treatments may be appropriate for very small eggs as found in molluscs but are inappropriate for fish eggs. The complex membranes of fish eggs present a barrier to penetration of chemicals but dechorionated eggs might be more permeable. Some success in the tench, *Tinca tinca*, was reported with intact eggs, however, by Linhart *et al.* (1986).

Timing and duration of treatments

The basic parameters of treatments for the production of meiotic diploid gynogenesis in a range of fish species are listed in Table 12.1. This range of techniques seems to be generally successful, and this form of breeding in fish is now more or less routine and dependent only on ancillary factors, such as ease of artificial fertilization and suitable incubation methods for the developing embryos.

In general, the exact timing of treatment after fertilization is not critical, but the window of opportunity is narrower at higher incubation temperatures. This is compatible with the cytological observations in gynogenetic eggs. The higher temperatures are also used for short periods and the duration of these is critical for survival of eggs – at low temperatures the duration of treatment is not critical.

Uses of meiotic gynogenesis

The use of gynogenesis, which involves the suppression of the formation of the second polar body, for chromosome mapping was covered in Chapter 7 and is currently one of its most useful functions (May and Wright, 1991).

Table 12.1 Pressure and thermal shock treatments for retention of 2nd polar body

Genus	Shock	Start m.a.f	Duration min	Reference
Pomoxis	36-40°C	5	1-5	Baldwin *et al.* (1990)
Pomoxis	5°C	5	45-60	Baldwin *et al.* (1990)
Pleuronectes	0°C	5-25	120-240	Purdom (1969)
Salmo	28°C	20	10	Johnstone (1985)
Ctenopharyngodon	2°C	1	5-15	Stanley and Sneed (1974)
Cyprinus	4°C	5-15	60	Nagy *et al.* (1978)
Oreochromis	4-11°C	15	60	Valenti (1975)
Oreochromis	42°C	2.5	3	Varadaraj (1990)
Misgurnus	-2°C	5	30	Oshiro (1987)
Misgurnus	500kg cm^{-2}	5	6-8	Vasetskii *et al.* (1984)
Oncorhynchus	28°C	10-50	10	Chourrout (1980)
Oncorhynchus	9000psi	10-50	10	Chourrout (1980)
Oncorhynchus	700kg cm^{-2}	5	6	Onozato (1984)
Carassius	3°C	2	20	Jiang (1982)
Oryzias	41°C	3	2	Naruse *et al.* (1985)
Brachydario	560kg cm^{-2}	3	5	Streisinger *et al.* (1981)
Ictalurus	5°C	5	60	Wolters *et al.* (1981)
Plecoglossus	0°C	6	60	Taniguchi *et al.* (1987)
Plecoglossus	600kg cm-2	6	6	Taniguchi *et al.* (1988)

Similarly its use in defining genetic sex determination is valuable. This was covered in Chapter 8 but some anomalies still remain for its application. The original idea, however, that the method could enhance the rate of inbreeding (Purdom, 1969) has largely been abandoned because the high level of crossing over now known to occur in fish chromosomes generates obligate heterozygosity in meiotic gynogenomes. The most promising avenue for the production of homozygous fish is mitotic gynogenesis which is covered in the next section. The use of diploid meiotic gynogenesis to generate clones following a cycle of mitotic gynogenesis is also covered in the next section. The principal benefit of meiotic gynogenesis in applied breeding now is its usefulness as an experimental approach to define methods for induced polyploidy (Section 12.4).

12.3 MITOTIC DIPLOID GYNOGENESIS

The mechanism here is depicted in Fig. 12.1 and entails the fusion of the products of the first mitotic division of a haploid 'zygote'. This simply brings together recently replicated chromosomes and therefore generates fully homozygous individuals at one step. It should be noted that the first generation of gynogenesis of this type produces individuals which are all uniquely homozygous, and that in order to produce a clone of such individuals a second cycle of gynogenesis is needed but this time it can be by the easier meiotic route.

Early attempts to induce mitotic gynogenesis in flatfish by applying cold shocks at the time of first mitosis in haploid embryos failed. Even though single-cell embryos were formed, they largely moved straight to the four-cell stage at the completion of the next cell cycle (Purdom, 1972b). Successful production of fully homozygous mitotic gynogenomes was reported, however, by Streisinger *et al.* (1981) using the zebrafish, *Brachydanio rerio*, a tropical aquarium species. The method used by Streisinger and his colleagues was to apply a combined treatment of pressure plus exposure to ethyl ether to eggs at the time of first mitosis after fertilization with UV-inactivated spermatozoa. Apart from the use of ether, which is probably optional, this was conventional technique but the genetic crosses were more complex than those used previously. Milt from wild-type males was used throughout and inactivated by UV irradiation. Females that provided eggs were homozygous for a recessive pigment deficiency gene, so that any inadvertant parental genetic contribution could be detected. In the first generation, pressure was applied during the period 22–28 minutes after fertilization, which is the period during which first mitotic metaphase occurs (the incubation temperature in this work was 27°C). Only about 4% of embryos survived to adulthood, and these included males and females even though presumed to be homozygous for all genes. These (different) individuals were crossed to produce an outbred group of offspring. Eggs from these were exposed to late-shock gynogenesis again to give individual females which were then put through the entire cycle yet again.

This was followed by a cycle of meiotic gynogenesis using heat shock to generate a clone of female embryos from each female of the previous mitotic cycle. Sex reversal of a proportion of these females enabled self-sustaining clones to be established and maintained thereafter.

The genetic basis for the need for this succession of three generations of mitotic gynogenesis is unclear and the presence of males and females in the offspring from them is a further genetic anomaly not yet resolved.

A repeat of this breeding scheme but without the multiple generations of mitotic gynogenesis was performed by Naruse *et al.* (1985) using a laboratory strain of the medaka, *Oryzias latipes*. The sequence was:

1. mitotic gynogenesis to generate unique individual females which are homozygous at all loci;
2. meiotic gynogenesis with individual females to produce clones of identical (isogenic) fish homozygous at all loci;
3. administration of methyltestosterone diets to some of these identical fish to create males;
4. mating of homozygous isogenic males and females to establish a stock.

No genetic confirmation of homozygosity was presented but survival of embryos to adulthood at 6% was not very different from the 4% recorded in the zebrafish. Where the two experiments differ, however, is that no males were found in medaka gynogenomes.

A third production of cloned fish by mitotic gynogenesis is reported by Komen *et al.* (1991) in the common carp. This breeding programme used the usual combinations of UV irradiation and heat shock but, in addition, employed a genetic system to monitor the presence of homozygosity. The females with which the breeding schedule began were heterozygous for one or both of a pair of genes which in the double recessive confer a blond colour to the fish. The females were also heterozygous for scale transparency. The female genotypes at the start were thus:

$$(a)\ \frac{b_1}{+} : \frac{b_2}{b_2} : \frac{t_p}{+}$$

$$(b)\ \frac{b_1}{+} : \frac{b_2}{+} : \frac{t_p}{+}$$

Meiotic diploid gynogenesis produces fish which are almost all heterozygous and therefore wild type in appearance. Mitotic gynogenesis from type (a) females produced 50% of each sort of homozygote so that four groups of offspring were found, i.e. homozygotes for both colour and scale transparency, homozygotes for one or the other and homozygotes for neither. There is an element of uncertainty about the latter, but not about the other categories. A single type (b) female produced 50% of the transparent scale phenotype and vice versa but only a low yield of double homozygotes for the other class as expected. The three genes are, of course, not linked.

Homozygous clones have thus been produced in two genera of cyprinids and one of cyprinodonts. The early attempts to achieve this in marine flatfish were unsuccessful and later attempts with salmonids seem inconclusive. Purdom *et al.* (1985) applied heat shocks 4.5–5.0 h hours after fertilization, and demonstrated homozygosity in rainbow trout by using three enzyme markers each of which individually gave high heterozygote frequencies following meiotic gynogenesis. However, 1 year after hatching none of the surviving fish were homozygotes, suggesting that the expected high level of 'inbreeding depression' had eliminated them. Other reports of homozygote production in salmonids are equally unsatisfactory. This does raise the question about the apparent low level of inbreeding depression observed in the carp, the zebrafish and the medaka. So far no follow-up work seems to have been reported with these pure inbred lines. The carp work was directed at understanding the genetics of fish immune systems and this appears to be a promising avenue of research. Whether the inbred lines can be put to practical use is not so clear. F_1 hybrids are attractive in most plant and animal breeding plans but as we have already seen (p. 75-76), isogenic 'ginbuna' carp did not behave like F_1 crosses, at least as far as growth rate was concerned.

One fundamental problem which might be addressed with gynogenetic homozygous clones is the nature of the haploid syndrome. Do the haploids generated by gynogenesis from the homozygotes still express these abnormalities, and if they do, how can the expression be explained?

12.4 ANDROGENESIS

The elimination of the egg genetic material, followed by fertilization and the development of the embryo from the chromosome set of the spermatozoa, is as old a concept as gynogenesis but nothing like so well understood. The difficulty partly resides in the fact that damage by ionizing radiation to the general contents of the egg is likely to be severe and to limit the subsequent viability of embryos. The advantage of diploid androgenetic eggs developed by fusion of the production of first mitosis is that the mechanisms cannot be confused with events involving meiosis. Many reports have been made of induced haploid androgenesis, but very few for diploid androgenesis in fish. Liu *et al.* (1987) report success in the loach, *Misgurnus anguillicaudatus*, but they apparently achieved this by mechanical removal of the female pronucleus followed by the insertion of haploid blastomere nuclei taken from elsewhere, and these duplicated spontaneously. Such techniques seem more appropriate to the cell genetics approach described at the end of Chapter 13, rather than to chromosome engineering practice.

Thorgaard *et al.* (1990) produced diploid androgenetic rainbow trout by the normal route and also by fertilizing γ-ray inactivated eggs with diploid spermatozoa from previously generated tetraploid males (Section 12.6). The comparison of the two methods showed clearly that fertilization by haploid spermatozoa followed by pressure shock at first mitosis was followed by much lower survival to hatching and to first feeding than the method using diploid spermatozoa. Thus it was concluded that the normally low survival of androgenetic diploids was a feature of the shock treatment, not the inactivation of the egg nor the genetic constitution of the embryos. With this increasing sophistication in the repertoire for manipulating chromosome sets, there is great opportunity for advancing knowledge on a range of polyploids and extending this to the phenomenon of the female-only parthenogenetic fish which generate eggs without going through a meiotic division cycle.

Further development of methods for induced androgenesis seems desirable, if only as a support to cell culture genetics or to manipulate male sex chromosomes for production, for example, of YY males.

12.5 INDUCED TRIPLOIDY

Current interest in induced polyploidy is almost entirely due to its potential application to fish farming. Conceptually it followed naturally from the work on gynogenesis, but using intact spermatozoa, not irradiated material, and the principal goal was to generate sterile fish (Purdom, 1972b). Prior to the impact of fish farming the main effort was with amphibia, but early work with fish was reported by Svardson (1945), Swarup (1959) and Vasetskii (1967). Even as late as 1972, Cuellar and Uyeno could report on the observation of a triploid rainbow trout without even an allusion to fish cultivation. Additional applied uses of polyploidy were the production of female-only fish (Stanley and Sneed, 1974) and the improvement of hybrid viability (Chevassus *et al.*, 1983). The impact of these applied aims has been to concentrate research much more on the practical aspects of induced polyploidy such as the means to recognize polyploids, the efficiency with which they can be produced and the direct advantages they have for use in aquaculture. None of these considerations was particularly important in most of the work on gynogenesis and androgenesis where the emphasis was on more fundamental and scientific aims.

Recognition of triploids

The mechanism for the induction of triploidy is the same as that for diploid meiotic gynogenesis (Fig. 12.2), but the recognition of the successful retention of the second polar body in triploidy is not so obvious as the contrast between haploids and diploids in gynogenesis. Several methods have been devised to assess whether or not a fish is triploid. There are no gross morphological differences, although minor differences have been observed between diploids and autotriploids (Bonar *et al.*, 1988), which have three sets of conspecific chromosomes. Allotriploids, which are of hybrid origin, do show morphometric variation simply because they comprise two sets of genes from one parent and one set from the other. They are, therefore, 'diluted' hybrids.

At the cellular level there are very clear differences between diploids and triploids of whatever origin. In the early amphibian work, and from the outset

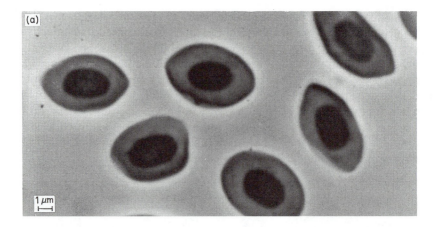

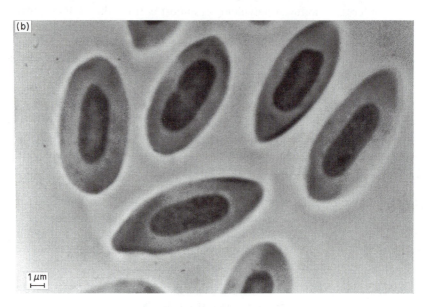

Fig. 12.4 Red blood cells of (a) diploid and (b) triploid rainbow trout.

of the fish work, the size of cells and their nuclei have been used reliably to identify triploids. The important aspect of this approach is to identify a specific and easily recognizable tissue. The commonest in use is the red blood cell (Fig. 12.4). Diagnosis of triploidy by red blood cell size is obvious at a glance, for an experienced operator, and does not even require measurement. A useful adaptation of this approach in fish by Johnstone and Lincoln (1986) for field use was to preserve small fry in formalin so as to be able at a later date to withdraw red blood cells from the heart to define the frequency of triploids in a batch. Another easily identified cell type, which can be seen in embryos before red blood cells can be extracted, is the cartilage cell in tissue around the cranium.

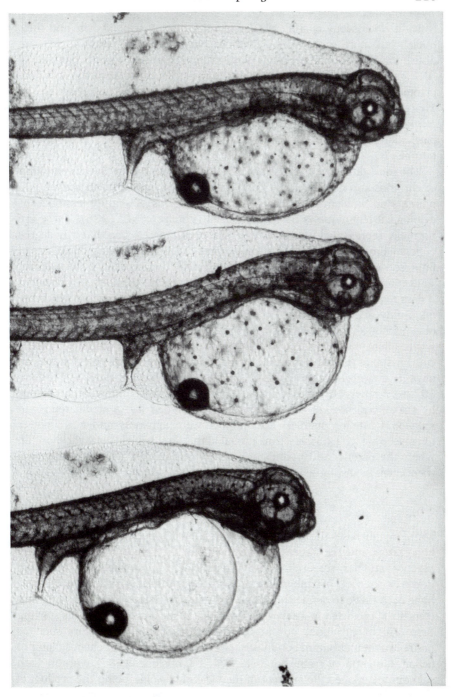

Fig. 12.5 Newly hatched larvae of the turbot (*Scophthalmus maximus*) (upper), its hybrid with the flounder *Platichthys flesus*, (lower) and the triploid hybrid (middle), showing pigment cell characteristics (TL about 10 mm).

Light microscopy is sufficient for identification of triploid cells and their nuclei, but more sophisticated methods have been used, such as cell size distribution by Coulter counter (Wattendorf, 1986) and various methods of DNA densitometry. The most popular of these electronic procedures is flow cytometry; its use with blood samples from rainbow trout was described by Thorgaard *et al.* (1982). Whilst being a bit too sophisticated for practical purposes, flow cytometry is likely to be of continuing research interest, particularly with the less obvious distinctions such as between triploids and tetraploids.

The most direct way to identify triploids is, of course, by the examination of chromosome number. This has been achieved in many fish species and presents no problem other than that it is laborious. A slightly less onerous cytological technique is to stain nucleoli (Phillips *et al.*, 1986). These are normally represented in cells as one per chromosome set so that any degree of polyploidy can be established by this method. Silver staining or the use of fluorescence microscopy as described in Chapter 7 seem likely to be useful.

Electrophoretic evidence of triploidy is also possible (Beck *et al.*, 1983) but it does require special circumstances such as triple allelic systems which are often only achievable with hybrids. As a research tool the technique is of obvious value but for practical purposes it too is laborious. Of greater simplicity where allotriploids are produced is the overall phenotypic or morphometric comparison of the hybrids and the parents. Thus larval pigmentation was an obvious diagnostic characteristic of triploids in marine flatfish (Purdom, 1972b). A bowl of triploids of plaice × flounder origin was immediately obvious by the appearance of the massed newly hatched fry, and differences in individual fish for pigment cell number were apparent under low-power microscopy (Fig. 12.5). It seems probable that any triploids of hybrid origin, where the parents and hybrids are easily distinguished phenotypically, will themselves be distinguished in the same way.

Yield of triploids

For many purposes the efficiency of production of polyploids is not of primary importance and the figures quoted in Table 12.1 are based on laboratory practice and for research use. For application in applied fish breeding, however, the yield of triploids assumes some economic importance and efforts have been made to define the conditions more precisely. The obvious parameters are those listed in Table 12.1, namely the form of the shock, its starting time and its duration, but in addition there are other factors such as pretreatment incubation conditions, egg quality – always an elusive factor to define – and imponderable hatchery factors. Johnstone (1985), working with salmon, introduced the concept of the yield of triploids, being the product of the survival of the eggs after treatment and the frequency of triploids in the survivors. He varied temperature from 26 to 32°C, duration of shock from 4 to 20 minutes timed from 0 to 35 minutes after fertilization, and achieved yields of triploids ranging from 0 to 70% and up to 100% triploid frequencies.

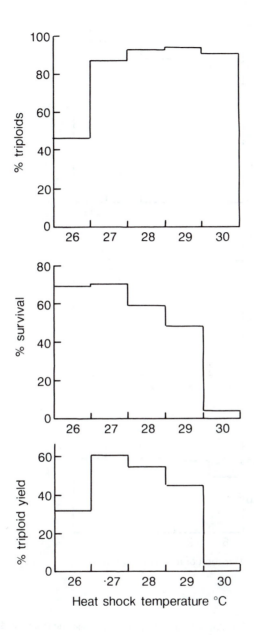

Fig. 12.6 The yield of triploids from rainbow trout eggs heat shocked at different temperatures.

The best yield was 70.5% following a temperature shock of 26°C for 20 minutes immediately after fertilization, but several were just below 70%, and the present consensus is that 28°C for 10 minutes starting 10 minutes after

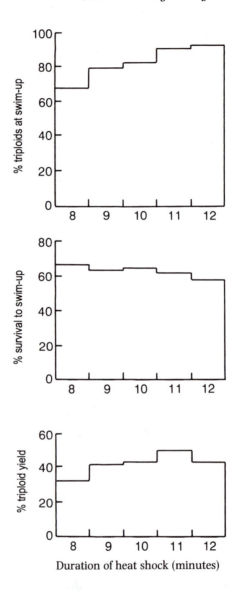

Fig. 12.7 The yield of triploids from rainbow trout eggs heat shocked at different durations.

fertilization is in the optimum range. A similar set of trials was conducted with rainbow trout eggs and the results are shown in Figs. 12.6 to 12.8. The optimal conditions for rainbow trout were similar to those for salmon.

Even with good quality eggs and with use of optimal conditions, there is still some loss of eggs following treatment with heat shock. This has led to the suggestion that pressure treatment might be less physiologically harmful than

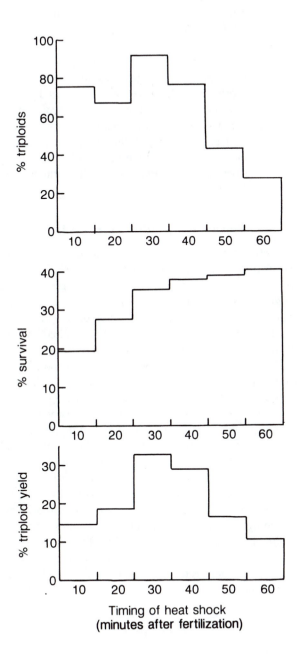

Fig. 12.8 The yield of triploids from rainbow trout eggs heat shocked for 10 minutes starting at different times after fertilization (note differing vertical scales).

high temperatures. Lou and Purdom (1984b) found no difference between pressure and heat treatment for rainbow trout eggs but Johnstone (1986) reports much better triploid yields in salmon eggs by pressure treatment than by heat treatment.

Commercial practice in the UK now is to use pressure treatments. For salmon farming, sterile triploids are of great value because they can remove the seasonality which otherwise characterizes the system. In addition, the use of triploids prevents the sharp decline of value of the crop which may occur if the fish start to become sexually mature before harvest.

The benefits of commercial implementation of new techniques have a cost. It is no longer possible to freely exchange information about improvements to methods – the information has monetary value and is now treated as 'commercial' in confidence.

Sexual maturation of triploids

The applied objective behind recent developments in induced triploidy was to generate sterile fish. That this was achieved was obvious from the first results, when triploid flatfish were shown to lack oocytes in the ovary whether from hybrid or straight cross origin (Purdom, 1972b). There was one complication, however, which involved differential gonadal development in the sexes. Although the theoretical expectation was that triploids would be sterile because of the disruption caused by three sets of chromosomes in meiosis, the exact nature of that sterility could not be predicted. It soon became apparent that in males the process of spermatogenesis was not seriously affected by triploidy (Lincoln, 1981a) and near-normal testes developed. The spermatozoa were not capable of producing viable zygotes, however, so sterility was confirmed at the genetic level. Similar studies confirmed that in the female triploid, ovarian development was almost entirely suppressed (Lincoln, 1981b). Exactly the same sequence was found in other triploids, e.g. rainbow trout, and the general conclusion is that male triploid fish are of little practical use in fish farming because they develop all the sexual characteristics of diploid fish. Females on the other hand do not develop functional ovaries and remain sexually indeterminate. This, allied to the methods described in Chapter 11 for sex ratio control, still provides a valuable breeding system for conventional fish farming. Where the sterile triploid male might find advantage, however, is in ocean ranching where return migration is essential, but sexual reproduction is not wanted.

Allotriploids

The use of hybrids in fish farming was described in Chapter 9. One of the common features of hybrids is that even though by combination of different parental attributes they may have potential value, their general lack of fitness is a problem. As a possible amelioration of this, Chevassus *et al.* (1983) proposed the use of triploid hybrids. There is little doubt that hybrid inviability is less in triploids than diploids. Distantly related species crosses are particularly affected by low viability, and improvements in triploids have been described in cyprinids (Liu, 1987; Matsunaga and Oshiro, 1987) and in salmonids (Arai, 1986), but these are of mainly academic interest. Similarly, survivors of distant hybridization may be triploids of spontaneous origin (Vasiliev *et al.*, 1975). Of more practical use, however, are the triploids in crosses between more closely related species. The tiger trout (*Salmo trutta* × *Salvelinus fontinalis*) is used in sport fisheries but has a low initial survival rate compared with either parent. Scheerer and Thorgaard (1987) described the production and recognition of triploid tiger trout and found that although they were better than the diploid hybrid, they were still less viable than the parent types. Quillet *et al.* (1987) examined a range of salmonid hybrids, allotriploids and autotriploids and concluded that the hybrid triploids had significantly lower survival early in their lives but compensated for this by better performance later.

Application of triploid techniques

Some comment has already been made on the application of triploid techniques to salmon farming in Northern Europe, and the extent to which this is increasingly covered by commercial considerations. Similar applications of the techniques to rainbow trout farming were pursued by government departments in the UK prior to this and were conducted in parallel with the sex ratio programmes described in Chapter 11. The surveys which were undertaken in connection with these sex ratio control programmes also embraced questions on the use of triploid techniques. These showed that most hatcheries used the method but that the overall use in terms of percentage egg production was much lower than that for all-female production and at about 30%.

Two reasons for this more restricted use of triploid rainbow trout were that triploid eggs were more expensive because of the loss during production and that their use was less important anyway, when the farm output was of portion-size fish at 1 year of age. The benefits of all-female triploids are only encountered if larger fish are desired and if the growth period on the farm or the survival time in a fishery are likely to extend to the period of normal sexual maturity in the fish.

12.6 INDUCED TETRAPLOIDY

The production of tetraploid fish follows the basic plan outlined for diploid mitotic gynogenesis but using normal spermatozoa; it is shown in Fig. 12.2.

Many of the earlier attempts to produce tetraploids were unsuccessful, but Chourrout (1984) and Myers *et al.* (1986) report success in rainbow trout. The tetraploids appear to be of very low viability, but nevertheless have been reared to sexual maturity and used to generate triploids by crossing with normal diploids, and androgenetic diploids in γ-irradiated eggs (Section 12.4). The difficulties still seem too large for routine use of tetraploids, but further developments with other species, for example those with lower chromosome numbers to start with, or with tetraploid gonadal transplant techniques, may achieve the necessary breakthrough. On the other hand, refining the existing methodology may well in itself improve the reliability of the tetraploid technique as has happened in other areas of induced polyploidy.

The facility to manipulate chromosome sets continues to have great potential for academic study of fish genetics and for practical application in a variety of commercial fish breeding activities, in which few attempts have been made so far to exploit these challenging opportunities.

Chapter thirteen

Gene manipulation

13.1 INTRODUCTION

Molecular biology has revolutionized the study of genetics in recent years as fundamentally as the discovery of Mendel's laws did in the last century. The structure and function of individual genes in a variety of organisms is largely understood, and the ways in which they replicate themselves and code for the construction of protein molecules is firmly established. Some organizational issues remain unclear, such as the significance of linear arrays of genes, the structure of eukaryote chromosomes and the function of the vast amount of trivial DNA, i.e. that DNA not involved in the transcription of proteins and frequently found as highly repeated simple sequences of nucleotides. Perhaps all of these mysteries are linked in one simple piece of biological logic.

Gene chemistry, the cutting of the DNA chain with specific enzymes and reassembling them with others, is well advanced and practicable in prokaryotes if not yet in eukaryotes. One exciting aspect of this chemistry is the possibility it has created of genetic engineering by gene manipulation. The objective of this is to insert desirable genes from one organism into the genomes of other organisms. It is now commonplace in microorganisms, achievable in a variety of higher organisms including vertebrates, and a highly fashionable pursuit at present in fish.

Gene transfer from one organism to another, or **transgenics** as it has become known, is very simple in concept – DNA is inserted into a nucleus so that it takes part in chromosomal replication and becomes part of the hereditary material of that cell. Although this is a simple idea in itself, the technology required to achieve it is very complex and in detail beyond the scope of this book. What follows is a brief indication of the procedures for gene manipulation by cloning and cutting and its application to gene transfer, but the reader is referred to Maniatis *et al.* (1982), Davis *et al.* (1986) and Sambrook *et al.* (1989) for detailed cover of these subjects.

The basic steps in the transgenic process are (i) deciding the appropriate gene and acquiring it in more or less pure form, (ii) injecting the material into a cell nucleus, following which (iii) its incorporation can be assessed by searching for the DNA itself as replicated by the host or for the RNA or for the protein for which it codes. Finally, the whole organism can be assessed for the extent to which the full expression of the gene occurs.

Before we consider the use of novel genes for transgenic work in fish, one early attempt at genetic transformation deserves comment. This is the wonderfully imaginative and well-constructed experiment by Vielkind *et al.* (1982) to introduce the gene controlling macromelanophore pigmentation in *Xiphophorus* hybrids into fish which lacked the gene. Following the pioneering work of Gordon *et al.* (1980) in gene transfer in mice embryos using purified DNA, Vielkind and his co-workers designed their experiment to use whole genomic DNA, i.e. DNA from the cell nucleus, but with a selective system of gene expression to facilitate recognition of successful transfer. The method involved extraction of DNA from fish showing gross melanotic tumour formation and the injection of it into fish lacking the tumour gene (Tu) but containing the various modifiers which enhance its expression. The injection was done in fish embryos and the DNA was placed in the vicinity of the neural crest, where pigment cells originate, at the time of rapid cell division. The expectation was that a Tu gene incorporated into a dividing cell would produce daughter cells with the gene and that this group of cells would be seen as a macromelanophore spot. The experiment was a success and transformation was claimed. Despite the elegance of the experiment, it still contained drawbacks not now encountered in gene manipulation using cloned DNA and molecular methods of assessment of gene location and expression. These drawbacks were that the DNA was not pure, the cellular expression could not be attributed to DNA incorporated chromosomally – DNA replication and expression can take place outside the nucleus – and finally, the cell systems were somatic and could not be tested genetically. Despite these limitations, the method seems sufficiently promising for a repeat, but using cloned DNA and molecular biological methods.

13.2 CHOOSING THE GENE

So-called 'gene libraries' for a range of animals, but especially for mammals, are arising in many laboratories around the world and much of the earlier work with fish transgenics used these mammalian genes. Fish gene libraries are now making progress and future fish work seems likely to concentrate more on these, but not necessarily at the total expense of other approaches, which will remain valid.

Two basic ways of gene cloning are used. The most direct involves the digestion of nuclear DNA by restriction enzymes (p. 99), the isolation of fragments and identification of their genic constituents following cloning by the insertion of the piece of DNA into a plasmid or virus, and the subsequent growth of these intracellular bacterial parasites to generate many copies of the section of DNA. The indirect method starts with tissue in which the gene of interest is functioning. The RNA is extracted from this tissue and predominantly represents that which is active in the transcription of the gene.

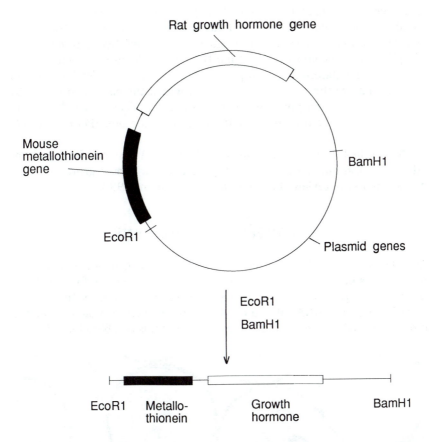

Fig. 13.1 Diagrammatic representation of the plasmid containing the growth hormone gene and its metallothionein promoter and of the linear molecule generated by enzymic cutting as used by Palmiter *et al.* (1982).

Following purification, the RNA is used as a template to form DNA by the use of mixtures of nucleotides and polymerase enzymes, and this reverse of transcription rebuilds the original gene sequence. This leads to production of complementary DNA (cDNA). The advantage of nuclear gene material is that it can comprise different parts of the gene, such as promoter sequences and natural spacers, which in themselves are not passed on during transcription. This appears to be important for eukaryotes but not for prokaryotes. The directness of the second method is of obvious advantage but the cDNA contains only those nucleotide sequences which code for the particular piece of RNA active in transcription of the protein molecule – the enabling bits are missing and for this reason it is necessary to use other promoter sequences to permit activation of the cDNA gene. These promoters are not specific so can be used for a variety of coding genes. For example, the plasmid used for the now-famous introduction of a rat growth hormone gene into the mouse

(Palmiter *et al.*, 1982) also included the promoter sequence for the mouse metallothionein gene. This plasmid has also been used for gene transfer to fish (Maclean *et al.*, 1987a) and is shown diagrammatically in Fig. 13.1. The plasmid itself comprises a circular DNA molecule into which have been spliced the metallothionein and growth hormone genes. Also indicated in the diagram are the locations of points at which the circle can be cut by specific restriction enzymes to generate a linear molecule, which appears to be more readily incorporated than the circular one (Penman *et al.*, 1990).

Most of the fish work so far has used cloned DNA designed for organisms other than fish but the range of fish gene libraries is now growing. These developments are reviewed by Maclean and Penman (1990).

13.3 INSERTING THE GENE

In order to achieve effective genetic transfer of a new gene it is necessary to insert copies of it into the cell nucleus at a time appropriate for the cloned DNA to take part in DNA synthesis. The normal method is therefore to introduce the new DNA into an embryo at the earliest possible time and as close as possible to the site of DNA replication. If this can be achieved before

Fig. 13.2 Two-cell stage in plaice eggs.

the first mitotic division of the fertilized egg then the incorporated material will be transmitted to all future cells. If one or more mitotic divisions occur before incorporation, a mosaic will arise with some cells containing the new gene, others not.

Fish eggs are ideal in most respects for this sort of work. They are available in large numbers, are subject to controllable external fertilization, are amenable to macroscopic handling and microinjection using standard micromanipulators, and are easy to incubate. The difficulty with fish eggs lies in the tough external membrane, the chorion, which surrounds them; this is a serious mechanical barrier to the use of delicate microinjection equipment but it can be overcome, either by chemical dechorionation using proteolytic enzymes (review, Hallerman *et al.*, 1988), which increases handling problems, or by puncture of the chorion with a solid needle prior to insertion of a microinjection needle. Retaining the chorion leaves the egg easier to handle after injection.

The DNA to be injected is suspended in an appropriate buffer (Penman *et al.*, 1990) and injected at the animal pole of the fertilized egg. Several hundred thousand copies of the cloned DNA molecule must be injected into each egg since a large number will be destroyed by cellular enzymes, and the probability that any one will be incorporated is very low and not dissimilar to specific mutation rates at about 10^{-5}. Because the single-cell state never becomes apparent in fish eggs – the first cells are those that form after the first mitotic division (Fig. 13.2) – the general aim to inject in the initial zygote nucleus is uncertain and the two-cell stage is sometimes chosen for treatment.

13.4 ASSESSING THE SUCCESS OF GENE TRANSFER

The earliest indication of the success of gene transfer into early embryos will arise during the period of intensive DNA replication which accompanies the rapid cell division stages of the early embryo. The joint replication of the introduced DNA alongside the recipient's own DNA can be detected in extracted DNA either by the use of a probe, a labelled molecule which specifically binds to a marker section of the introduced DNA (Southern, 1975) and can itself be detected for example by autoradiography, or by the digestion of the extracted DNA with restriction enzymes followed by electrophoretic separation to demonstrate the presence of the unique sequence of the introduced molecule. Both methods taken together are convincing proof of incorporation, and if the probe label moves with high molecular weight DNA in the absence of cutting by specific restriction enzyme, this is evidence of an association between genomic DNA and the introduced molecule.

Zhu *et al.* (1985) injected a bovine papilloma virus construct containing a genomic human growth hormone sequence and the promoter for mouse metallothionein into dechorionated eggs of the goldfish, *Carassius auratus*, at the pronucleus stage. The circular plasmid genome was characterized by restriction sites as shown in Fig. 13.3 and injected as the linear sequence

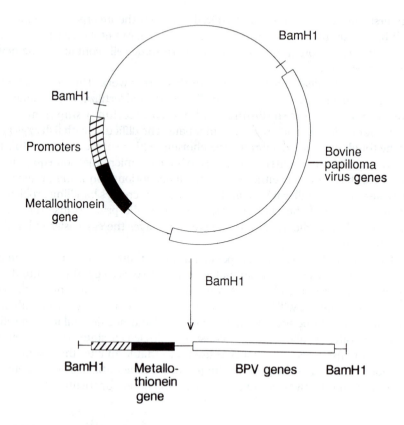

Fig. 13.3 The novel gene sequence used by Zhu *et al.* (1985) and the plasmid from which it was derived.

produced by cutting this with the BamH1 restriction enzyme. The linear construct included a segment 2.3 kilobases long of the original bovine papilloma virus (BPV) molecule, for which a ^{32}P labelled probe was available, followed by a DNA fragment containing the mouse metallothionein promoter and the human growth hormone coding sequence and flanking sequences.

From DNA extracted from embryos killed at successive stages during embryogenesis it was shown that much initial replication of the introduced molecule occurred after it had undergone restitution to the circular form and that this rapidly diminished in later embryonic life. Remaining labelled DNA was not of the circular form, however, but appeared linked to the higher molecular weight fractions of the extracted DNA. In other words the incorporation of the probed bovine papilloma virus sequence was demonstrated and it was assumed that this also applied to the rest of the original molecule. The analysis of DNA from fish at 50 days of age showed that the new gene could be detected in about half of them, that it migrated electrophoretically with high molecular weight DNA and that it could be identified at its original molecular weight from DNA cut with the restriction enzyme BamH1. Similar

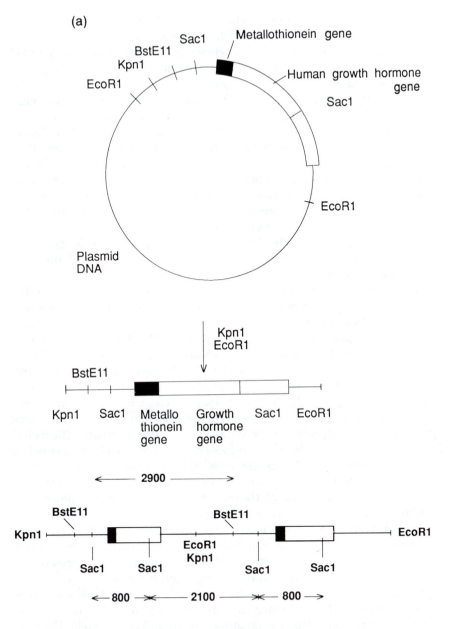

Fig. 13.4 (a) The plasmid and derived linear model used by Dunham *et al.* (1987) in channel catfish. (b) The structure of the incorporated sequence as revealed by cutting with the restriction enzymes Kpn1, BstE11, Sac1 and EcoR1.

demonstrations of the transfer of growth hormone genes into rainbow trout embryos have been described by Chourrout *et al.* (1986) and Maclean *et al.* (1987b), but using different DNA constructs and slightly different methods of injection. Chourrout *et al.* employed a sequence containing a cDNA version of the human growth hormone gene and injected it into the cytoplasm of the animal pole of the trout eggs at 2–6 h after fertilization.

Maclean *et al.* used the construct employed successfully by Palmiter *et al.* in mice (Fig. 13.1), which included a genomic version of the rat growth hormone gene, injected directly through the chorion into the two-cell stage of the embryo. Penman *et al.* (1990) examined various factors which might influence the efficiency of gene transfer within this system and confirmed that the linear molecule was incorporated more readily than the intact circular version, as observed in mammalian work. They also demonstrated that there was a degree of dose response following injection of varying numbers of DNA molecules but that the nature of the buffer used to suspend the molecules and the method of injection had no effect. The techniques for achieving integration of novel genes into fish, however, are still at an early phase of development and much improvement seems likely in future.

A further construct (Fig. 13.4) was used by Dunham *et al.* (1987) to transfer a metallothionein–human growth hormone gene into channel catfish, *Ictalurus punctatus*. The germinal disc of fertilized eggs was injected at times ranging from 5 minutes to 90 minutes after fertilization at which time the first cleavage took place. Fish were killed 3 weeks after hatching, and extracted DNA was examined for incorporated parts of the injected linear molecule. A gene probe was available to detect that fragment between the restriction sites Kpn1 and EcoR1, indicated in Fig. 13.4(a), and further digestion by restriction enzymes Sac1 and BstE11 was undertaken to clarify the structure of the incorporated DNA. Digestion of fish DNA by Sac1 generated a fragment 2100 kb units long which included the Kpn1/EcoR1 probed bit and could only be explained on the basis that the incorporated gene was a tandem duplicate as shown in Fig. 13.4(b). Digestion with BstE11 produced the 2900 kb fragment to confirm the tandem duplication structure of the gene insert. Another mouse metallothionein–human growth hormone fusion gene has been incorporated into eggs of tilapia, *Oreochromis niloticus*, this time via injection through the micropyle (Brem *et al.*, 1988), and verified in much the same way as in channel catfish.

Several other genes have been incorporated into fish. Chicken crystalin, a simple eye lens protein, was transferred to *Oryzias latipes* (Ozato *et al.*, 1986) and shown to be expressed. The metallothionein gene itself, as opposed to its promoter sequence mentioned earlier, has been incorporated into rainbow trout embryos (Maclean *et al.*, 1987b) and could have significance in conferring resistance to heavy metal pollution. Similarly purposeful is the introduction of a gene controlling the production of an antifreeze protein (FSP) into salmon (Fletcher *et al.*, 1988). The objective of this was to develop a strain of salmon which would resist sub-zero temperatures in Canadian waters. The work entailed the use of a cloned fish gene, that which codes for FSP in winter flounder, *Pseudopleuronectes americana*.

In a further step forward, Yoon *et al.* (1990) developed methods for transferring a bacterial gene conferring neomycin resistance into the goldfish, *Carassius auratus*. The objective of this was to move towards a system of self-selection for transformed genotypes. Thus if a large number of cells or organisms is treated to generate only a low frequency of transformed units, the use of a selective agent to isolate the latter is highly desirable. If the neomycin resistance gene is joined to the definitive gene for which transfer is sought, e.g growth hormone, then by treating all the exposed individuals to neomycin all but those in which transfer was successful will be killed.

There is much proliferation of successful transfers. Less obvious is successful expression of the transferred gene.

13.5 EXPRESSION OF TRANSFERRED GENES

The enormous obese-like mice produced by Palmiter *et al.* (1982) following the transfer and expression of rat growth hormone genes caught the popular imagination, but expression in fish of transferred genes is the exception rather than the rule. Most of the published work stops short of describing expression but there does appear to be a problem (Maclean *et al.*, 1987b). It has been suggested that cloned fish genomic DNA is the best material to use to achieve expression, in that the artificiality of the cDNA molecule and the possible inappropriateness of supportive DNA sequences (promoters, spacers etc.) from non-fish species make good expression in the host unlikely. The second part of this argument seems too simplistic, but several laboratories around the world are working on fish genomic libraries so that this idea can be put to a wide test. In the meantime, some slight evidence of expression exists for the crystalin gene (Ozato *et al.*, 1986), the metallothionein gene in embryos (Rokkones *et al.*, 1989), and the neomycin resistant gene (Yoon *et al.*, 1990), but none of this is wholly convincing.

Part of the problem of expression no doubt lies in the actual composition of the inserted sequences in terms not only of which genes are involved but in what order. In addition, differential cell expression is a common feature of simple protein production in multicellular organisms. There would be no point, for example, in introducing a gene for amylase production in fish – to make them genuinely vegetarian – if the enzyme was produced in the brain as opposed to, say, the pancreas. A further point is that pleiotropic interactions of genes are common and together with sundry modifying genes make for subtle control of the form and function of the whole organism. It is in this context that the apparent abnormality of the giant mice should be viewed – they were not just big mice. The dramatic consequences of the appearance of a new protein or of significantly greater quantities of an existing one may therefore not elicit a standard response. Growth hormone is a good example of possibly conflicting aims. Much experimental work has been done with growth hormone administered by feeding or by injection. Recent studies by Agellon *et al.* (1988) and Down *et al.* (1989) claim large effects of such hormones injected into salmonids but others have been negative, particularly those using diet trials. Danzmann *et al.* (1990) found no effect of growth hormone on growth in weight although an effect on fish shape was recorded.

Finally, Scott and Power (personal communication) find significantly greater growth of those fish which received growth hormone injections in a population of treated and untreated rainbow trout. The problem is that growth is controlled by appetite, food availability, social hierarchies and other factors, and to change one without the others being adjusted may achieve nothing.

Further work on the expression of novel genes transferred to fish is clearly needed before this important branch of modern genetics can progress further. It seems likely to benefit from the widening range of fish genomic libraries but new concepts may be necessary to generate further advance.

13.6 FUTURE DEVELOPMENTS IN MOLECULAR GENETICS

The injection of large amounts of DNA into the cytoplasm or pronucleus of an egg remains a crude hit-or-miss activity in which the whole organism must participate. Testing for expression is a lengthy and, up to now, not very rewarding exercise, and demonstration of truly genetic incorporation, i.e. into germ cells, is limited in fish of commercial importance by their long generation time.

There is a need for a more streamlined approach, and something along the lines of tissue culture genetics seems appropriate. A growing interest in biology is the development of stem cell tissue cultures and even the growing of totipotent cells, such as egg blastomeres, in culture. Fish could provide excellent opportunities for such work with the possible development of haploid or diploid systems by gynogenesis. Thus the production of cells containing new gene arrangements could be undertaken using tissue culture genetic techniques and screening for useful or interesting combinations. Insertion of nuclei from such cultures into enucleated or androgenic eggs would provide a means of generating whole fish containing the new genes.

Along with stem cell culture the next phase of molecular genetics will need to examine the causes of cell differentiation, including changes to nuclear DNA, and the possibility of reversing them, thus making it feasible to take any dividing cell from an adult organism to create nuclei which could be inserted into eggs by micromanipulation to generate clones. Enormous ethical problems can be anticipated from such advances. Fish would appear to be ideally suited to this type of biological research from technical and moral standpoints.

Chapter fourteen

Featured fish: I. Ornamental species

14.1 INTRODUCTION

Fish from a wide range of taxonomic groups have been kept in captivity for their aesthetic appeal. Sterba (1954) provides a comprehensive review of these together with details of their systematics, distribution and basic needs in captivity. Relatively few of these species have been bred in captivity, but it remains a common goal of amateur aquarists to breed the more difficult types of fish. Of those successfully bred, a mere handful have been subjected to domestication involving genetic modification. Of these, the oldest is without doubt the goldfish whilst the koi has gained enormous popularity over the past few decades. By far the most widely bred fish and those most intensively modified by genetic selection, however, are a few species within the family Cyprinodontidae of which the Guppy and the platy provide the obvious examples. This small group of fish is revered by aquarists all over the world for its exotic appeal. The only other species widely bred for its colour and other features is the Siamese fighting fish, *Betta splendens*, of the family Anabantidae.

14.2 THE GOLDFISH

The goldfish is a domesticated form of the so called Prussian or silver carp, *Carassius auratus*, which is still widely distributed in Asia as well as in Eastern Europe. The wild fish are generally of a silvery appearance but may also be of a brassy hue, but nothing like the red-orange appearance of the domesticated goldfish. This coloration and other aspects of present-day goldfish are believed to have been developed from the wild type, somewhere in China, before the 16th Century. They are reported to have been introduced into Japan in the 16th Century and into Europe about 100 years later. Various explanations for their quite distinctive coloration have been given of which the most rational are selective breeding from albino stock, distant hybridization and

Fig. 14.1 Fancy strains of the goldfish: veiltail (upper) and comet (lower) (TL about 14 cm).

degradation of pigment cells by virus particles. Crossbreeding does not support the first postulation since intermediates are not observed, and no evidence of viral replication supports the third, but some evidence in support of distant hybridization was given in Chapter 9. Whatever the genetic basis, the gold colour appears during the early months of life of the fish and seems to involve the progressive depigmentation or destruction of melanophores. The brilliance of the final colour is probably determined environmentally. The goldfish of warmer climates, Israel or India, for example, certainly seem of a vastly more colourful appearance than those reared in the cooler temperate regions of Europe.

There are three basic body colours in goldfish varieties. The commonest is the rich red-gold appearance, but all-black is typical of some forms and the well-known shubunkin exhibits a blue, red and black mottled appearance plus the transparent scale characteristic detailed in Chapter 4. Kirpichnikov (1981) reviewed evidence on the genetic control of these major colour variants. Blue is reported to be determined by a simple Mendelian recessive with brown coloration under more complex control. Depigmentation itself is described as under control of two dominant loci, the recessive alleles of which confer the black coloration (Kajishima, 1977). It seems very likely that these genetic models are oversimplified and that the ancient origin and long history of selective development have generated a highly complex genetic background for body colour.

The fin and body shapes of goldfish (Fig. 14.1) are also of great antiquity but have undergone much recent development. The range of varieties is very large – including body colour, there are reputed to be almost 200 individual strains of goldfish, i.e. forms for which body proportions and colours have been defined for the purpose of selective breeding for show purposes where competitive judging takes place. The probable major gene/polygene control of the forms of these fish was discussed in Chapter 4. Selective breeding is likely to be feasible for modification of such fin and body shape traits. Selection by culling can be very efficient where visible criteria are employed and where fecundity is large as in most fish species. The combination of the use of such selection to amplify possible effects of major genes which arise fortuitously in breeding programmes was discussed in Chapter 6. When performed by an army of amateur and professional breeders around the world, within the discipline of agreed show standards, such selection easily encompasses the large phenotypic variety in goldfish and, indeed, in most other domesticated animals and plants.

New approaches in a fish like the goldfish with such a distinguished history of selection seem almost presumptuous but one possibility is to attempt crossbreeding with common carp forms to transfer the 'mirror' and 'line' phenotypes into the goldfish. Such developments have taken place in koi – which is, of course, the fancy variety of common carp. The essential features of goldfish are for viewing side on – i.e. it is an aquarium fish more than a pond fish. The opposite is true for koi. The development of scale patterns might therefore add a new dimension to the traditional goldfish breeder's repertoire.

14.3 KOI – ORNAMENTAL COMMON CARP

The name koi in Japanese means carp but these are usually taken to be the colourful derivatives of *Cyprinus carpio*, alternatively called nishikigoi (coloured cloth in Japanese), and have a degree of antiquity similar to that reported for the goldfish. Developed in Japan from black ancestors originating in China, the coloured varieties date from the 16th Century. In contrast to the situation in goldfish, however, the major colour and scale variations popular today have been generated in the last 50 years and are still being added to. These fish enjoy a world-wide reputation and command extraordinarily high prices from devotees and those wishing to display such major treasures. Like goldfish, they can live for many years and represent valuable 'investments'.

Present-day koi varieties number over 100 but a basic set of 13 is used for judging fish in shows. The show varieties comprise combinations of solid body colour and blotchiness involving red, white, blue, black, yellow and orange coloration plus variations in scalation.

They are classified according to a Japanese system devised for show purposes and the names frequently reflect aspects of the varieties' history or place of origin. A subsidiary classification of a more descriptive type overlays

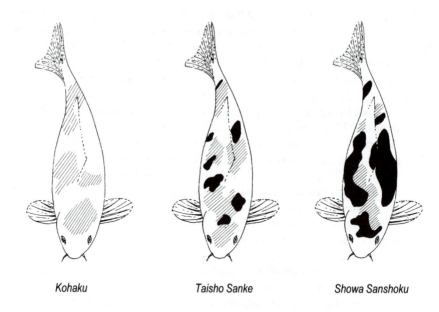

Kohaku   Taisho Sanke   Showa Sanshoku

Fig. 14.2 The three major koi show varieties with red (hatched) or red and black (solid)
  patches on a white background (TL up to 60 cm).

the traditional one to give a binary system of great complexity comprehensively reviewed by Tamadachi (1990).

Simple patches of one colour superimposed on the background of
another (Fig. 14.2) is the basis of most of the traditional varieties. Thus
Kohaku are white fish with red patches, *Taisho Sanke* in addition have
smaller patches of black within the red and white whilst *Showa Sanshoku*
contain much more black, to the extent that it can be regarded as a black
fish with added colour. The aristocrat of these varieties is the *Tancho
Kohaku*, a white fish with a solitary red patch on top of the head: in the
most valued specimens this is round like the rising sun symbol on the
national flag of Japan. All of these varieties, plus the non-valued all-white
specimens, are said to be found amongst offspring of any cross within these
varieties.

Further varieties include black on white (*Bekko*) and its obverse (*Utsuri*)
shown in Fig. 14.3 together with *Shusui* in which linear scale patterns are
combined with other characteristics. These scale patterns are closely akin to
the scaly and line characteristics of common carp and probably derive from
a similar pair of Mendelian characters as described in Chapter 4.

Colour patterns are not so simply inherited but may also be due to the
varied segregation of a few simple Mendelian alleles. As described in Chapter
3, such characteristics can generate an almost bewildering array of phenotypes. There is also a common theme in fish genetics of duplicate loci. Where
this can lead to tetrasomic inheritance, i.e. four genes segregating not two,
even greater complexity arises.

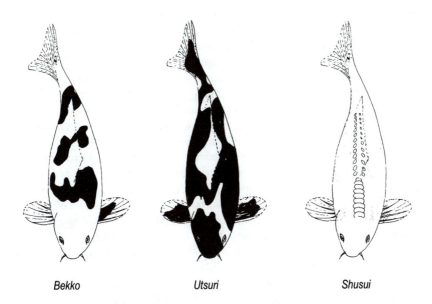

Bekko   Utsuri   Shusui

Fig. 14.3 Three further koi varieties which are popular for show purposes. *Bekko* and *Utsuri* can involve black (solid) patches with either white yellow or red elsewhere. *Shusui* combines blue coloration in the linear array of scales with red and white body colour.

colour. Other body colours of orange and yellow are also found, and a blue type probably relates to a dilution of the normal black pigment. Scale pigmentation variation characterizes some varieties, and others show the metallic sheen probably produced by excessive deposition of guanine in the scales.

The history of the varied colour patterns is not well documented but seems to follow a pattern of inbreeding with the chance occurrence of distinctive mutations. Little of Mendelian type analysis has been done within koi, but in a series of reports Katasonov (1973, 1974, 1978) attempts to define the genetics of some of the colour elements from crosses of ornamental fish with wild-type common carp. The overall gold form is stated to be the consequence of recessive alleles at two loci whilst the metallic sheen is due to one single recessive allele. All three in combination produce the white fish. In addition to this it seems likely that scale transparency follows the model observed earlier with common carp and is determined by a recessive or partially dominant gene. The reticulation pattern of melanophores on scales resembles 'gold' in the Guppy and that was controlled by a simple recessive allele. Thus far seven loci seem to be involved, with differing degrees of dominance, and the various patterned phenotypes seem likely to embrace a few more. The genetic situation with respect to koi varieties is thus very complex: strains breed true to some basic features but give rise also to much further variation, and the prospects for developing yet more forms seem excellent, particularly if different strains are crossed – what such practice is likely to do, however, is to generate varieties are crossed – what such practice is likely to do, however, is to generate varieties which fall outside currently accepted show standards. As with goldfish and, indeed, all animals and plants grown for show purposes, beauty is in the eye of the beholder or, in this case, the judges.

Most of the world's finest koi now come from Japan but a growing interest in selective breeding of these fish is taking place in many countries. Import legislation as a means of combating the transfer of disease is also growing and this may well enhance the need for 'home grown' fish of the koi varieties.

New developments include the introduction of fin shape into the koi gene pool, and body shape may come next. The flowing fin varieties of koi are called hifin veiltail koi, presumably after their counterparts in the platy. They appear to be only of the simpler body colour types and in this respect follow the pattern to be seen with the linear scale types. There is no documentation, but it might be presumed that these scale and fin derivatives of koi have originated recently by crossing of ornamental koi with carp or goldfish followed by intensive backcrossing.

Chromosome engineering would also be usefully developed in koi culture if only as a means of protecting valuable genetic stock by providing for sale only sterile triploids. Thus the selective breeding of koi has some potential but is likely to be constrained by show-judging specifications. What is also clear is that the finer points of those individual fish judged to be superior are unlikely to be inherited but most probably are of chance, environmental origin. Were this not so, the huge value placed on such perfect specimens would be eroded by mass production.

14.4 THE TOOTHCARPS

Whereas koi and goldfish are basically outdoor fish, and although the latter can live well enough in small aquaria, the small colourful tropical fish of the family Cyprinodontidae are ideally suited to small indoor aquaria. They make excellent subjects for developmental breeding because of their short life histories, ease of rearing and natural propensity to phenotypic variation. Much has already been made of the genetics of pigment characteristics for the Guppy and the platy in Chapter 3, and the variable pigment patterns in nature continue to provide material for scientific study (Borowski, 1984; Shaw *et al.*, 1991). The artificial breeding of both of these species has, however, taken variation a lot further forward than that found in the wild.

The Guppy

The natural variation in these fish comprised dots of colour and black pigmentation and some extensions of the caudal fin, mostly sex linked in inheritance and limited in expression to the males. The principal developments in fancy Guppy breeds include the production of distinctive body colours and very greatly extended caudal fins as in the veiltail Guppy. The body colours red, blue and green seem likely to be derived from the inheritance of simple

Mendelian characteristics of autosomal linkage. The black Guppy, however, is probably derived by selection of the Half-black phenotype due to a gene located in the X chromosome and having a dominant effect in both males and females. The most obviously selective improved characteristic of these fish, however, is the greatly extended caudal fin accompanied on occasion by elongation of the other non-paired fins. The complex genetics of this has been described by Schröder (1974) but its greatest impact seems to be when it is developed with other genetic features, such as the lace tail gene giving a yellow filigree appearance to the enlarged caudal fin, Half-black, and Flavus, a brilliant yellow-bordered fin, or the solid body colours of red, blue or green. The overall genetic control mechanisms are simple, but the combinations are very numerous and the range of strains which have been produced is a tribute to the skill and patience of amateur breeders. Further developments seem assured; one special goal should be the separation of the factors determining size of the paired and unpaired fins.

The platy

Fish of the genus *Xiphophorus* are amongst the most variable of animals within the natural environment. They also comprise a range of species which hybridize readily amongst themselves and thus provide the breeder with two sources of variation with which to work.

Three sets of colour are inherited: the macromelanophore and micromelanophore patterns so valuable to scientific study (Chapter 3) and a series of body ground colour or fin colours. In addition to this, fin shapes and sizes are also developed in the fancy strains. The genetics of the fin shape may derive from hybridization between *X. maculatus*, the common platy of aquarium fame, and *X. montezumae* or *X. xiphidium* followed by selection for enhanced expression. The other characteristics have a complex genetic background involving at least 19 loci (Kallman, 1975), but these too seem to have been selected for vivid expression of individual genes. The possible manifestations of this array of individual loci are very numerous, and the practical problems of platy breeding are largely to do with keeping crosses simple and stabilizing the variation. The black and black-stippled forms probably represent another example of duplicate locus inheritance.

These fish make excellent material for educational use. Some of the simpler strains such as the moon platy are less decorative than the more vivid fish and certainly have a less complex genetic background than present-day Guppy stocks. Crossbreeding with the readily available swordtail also provides good material for demonstrating hybridization, and the melanic forms deriving from this show a good response to selection.

14.5 OTHER SPECIES

The widely distributed diminutive threespine stickleback, *Gasterosteus aculeatus*, should provide good material for genetics and selective breeding.

Its scale characteristics have an interesting genetic background which is by no means fully understood. In addition, it displays other polymorphisms of finnage and colour which have not been studied genetically. Its ability to adapt to large lakes and reservoirs, to colonize small waterways like drains and ditches, and to be successful in fresh or salt water make it suitable for ecological genetic research.

The biology of sticklebacks is well understood (Wootton, 1976) and breeding and rearing are not too difficult (Bell, 1984). Finally, unlike the present situation for most of the species described so far in this chapter, the stickleback can be obtained in wild-type form easily, from most bodies of unpolluted water in the northern part of the Northern Hemisphere.

Chapter fifteen

Featured fish: II. Cultivated species

15.1 INTRODUCTION

Fish cultivation is an old practice but has grown in importance in the last 20 years to the point where it generates almost one-tenth of the world's fish harvest. In addition the production of fish to supplement or replace the natural species in sport fisheries has been stimulated by increasing demand, particularly in the more densely populated regions of the developed nations of the world. This encompasses the breeding or rearing of many species of freshwater fish and a few marine examples. Of these, the ones already developed in the sense of genetic modification are carp, salmon and trout of various species, catfish and tilapias. Apart from the common carp, the others have all emerged as potential domesticants over the last 100 years and, compared with other domesticated vertebrates, are still very close to their wild ancestors. There is therefore great potential for genetic modification.

15.2 COMMON CARP

The best-established farm fish and a member of the family Cyprinidae, *Cyprinus carpio* is farmed in temperate and tropical regions all over the world and is revered as a sport fish by some and hated as a foreign intruder by others. It is extremely hardy, will grow at temperatures in excess of 8°C but requires hot summers for natural reproduction. Despite centuries of culture by Man, as a farmed fish it is largely unchanged from the wild type, except for the development of the scale phenotypes described in Chapter 4 and possibly the divergence of body proportion exemplified by the contrast of Asian and European forms discussed in Chapter 5.

There does not seem to be any major reason for selective improvement in the general features of this species. Growth rate is not responsive to selection but since these fish are grown in extensive culture systems where temperature, stock density and food availability are the principal determinants of growth, there is probably little call for genetic improvement in this context. Specific improvements such as reduction in intramuscular bones have been attempted without success. These needle-sharp bones lie free between the muscle myotomes and cannot be removed by normal filleting – nevertheless, a solution through processing seems more attractive than a genetic answer although in many parts of the world the fish are marketed alive. Of greater relevance, perhaps, would be development of flesh colour or texture since there are signs that demand for this species is not always high. Scaliness too has commercial relevance. The inherent problem of recessive lethality in the 'leather' carp (p. 45) might be avoided if 'mirror' carp were selected for reduced scale expression. Some evidence for genetic variance for this trait was mentioned (p. 46).

The greatest potential for the genetic improvement of carp must lie within the field of sex ratio control and possibly in sterility in induced triploidy. This species has relatives within which all-female types can arise, and hybrid crossing to achieve this is of interest although success cannot be predicted. Selection for specific sex ratio traits does seem feasible, however, in view of the complexity of sex determination in the species as revealed by gynogenesis.

Further development by chromosome engineering seems attractive. Triploids have been produced but a major event would be the production of all-female triploids which could reproduce by gynogenesis. Triploid hybrids with distantly related species might be the starting point for this development. The cyprinid scope for all-female diploid forms should be followed up more fully. A broodstock of predominantly female form to generate triploids, either by standard heat shock or by crossing with males that generate diploid spermatozoa, seems a not unlikely development. The production of diploid spermatozoa can be achieved either from tetraploid fish or from males arising as sex-reversed individuals of the type possible in gynogenetic silver carp.

Where carp are pests, a genetic programme to introduce sterile triploid males could have an impact similar to that achieved in the control of screw worm. Triploid males will generate spermatozoa and fertilize eggs, but the zygotes will all be inviable.

15.3 SALMON AND TROUT

The family Salmonidae is well represented in the cultivated fish list. Several species of Pacific salmon of the genus *Oncorhynchus* together with Atlantic salmon, *Salmo salar*, are now cultivated. In addition, a range of trout species is farmed. The most ubiquitous is the rainbow trout, *Oncorhynchus mykiss*, found in virtually all temperate regions of the world, but brown trout, *Salmo trutta*, and brook trout, *Salvelinus fontinalis*, are also popular species, particularly for use in sport fisheries.

A major feature of these salmonids is that they are all subdivided into numerous races or strains. This is a consequence of two basic features of their life styles. The first feature is that many of the species exhibit an anadromous life history in which they return very precisely to the rivers or even the tributaries of their birth, thus paving the way for development of a complex intraspecific stock structure. A similar phenomenon arises with lake-dwelling species which migrate to running water to spawn. The second feature is the existence of many land-locked populations in which genetic divergence has occurred because of the absence of opportunities for the mixing of populations and the genetic conformity this would bring. The barriers to this gene flow may be physical, such as the isolation of lakes left behind by receding ice ages or the presence of waterfalls on rivers. They can also be biological, in that two populations living together for part or the whole of their lives are separated by spawning, migration or other behavioural characteristics.

A further course of divergence is, of course, domestication. Rainbow trout and brown trout have been reared in closed populations for over 100 years. Whether the current strain differences are simply due to divergence during the period of domestication or, more likely, arose because different natural populations provided the starting stock, is immaterial. A wide range of strains or geographic stocks is available for most of the salmonids and these should be exploited before expensive and unreliable selection programmes are initiated.

Growth rate is an obviously important characteristic in a farmed species and wide natural ranges exist in this context. Spawning time is also highly variable, particularly in rainbow trout, but also in other species, and this too has implications for fish cultural practice. More complex is the sexual life history of salmon which may encompass maturity at the parr stages, 1 year later as jacks or grilse or 2 or more years later as salmon. These mature fish may also enter their spawning rivers just prior to spawning time or they may opt for early migration up to 1 year before the actual spawning event. The genetic control of these complexities is not known in any detail but they are of very considerable interest from both the natural and fish culture points of view.

Much less attention has been given to the use of specific strains of fish for sport purposes, and introductions have often been of unspecified, probably unknown stock. That this is poor management per se seems undeniable but it also may be contrary to conservational aims if an introduced stock can compete with another. Of direct importance is the choice of fish for the fishery itself. Thus it is common in brown trout fisheries to take stock from a river to rear in captivity so as to return grown-on fish to the same river at a later date, possibly as progressive supplements to an overexploited resource. Even this approach has been criticized on conservational grounds as a threat to the genetic integrity of stocks, but this seems an unlikely source of drastic change.

More complex choices can be made within fisheries supported totally by introductions and where no contact is made with significant natural populations. The use of rainbow trout itself is the basic example of this management tactic. These fish are undeniably the easiest of trout to rear and the fastest growers. It is still possible within rainbow trout to make specific choice of a variety of hatchery strains which differ in colour characteristics, spawning behaviour and growth rate. One of the more fundamental choices which could be made is between anadromous (steelhead) and non-anadromous (resident) rainbow trout. The former are more active and more salmon-like in appearance than the latter and by the use of hybrids, a combination of features can be chosen.

Wide choice is also available in brown trout. The distinctive natural populations of these fish offer choices of feeding habit. Some are basically demersal, others more pelagic in their feeding patterns, but perhaps the most obvious distinction is between the ferox type of fish-eating habit and the more normal invertebrate consumers. In many artificial fisheries the natural reproduction of cyprinid fish presents not only a ready source of prey but also a significant competitor to trout in that they too feed on planktonic animals. Use of ferox-type brown trout could exercise a degree of control on such populations.

The most promising modern developments for salmonid farming are without doubt the use of female-only stock and sterile triploids. These subjects were covered in Chapters 11 and 12 but their obvious relevance to current salmon farming practice should be emphasized. One of the greatest difficulties in salmon farming is the seasonal nature of production methods. Size is important in that the fish are grown for as long as possible and harvesting takes place just prior to sexual maturity in the autumn. Care is also needed to avoid early maturity, either of parr or of grilse in the sea cages. All of these difficulties would be avoided by using triploid females, and a regular cycle of production year-round could be achieved.

15.4 TILAPIAS

These fish of the genera *Oreochromis, Sarotherodon* and *Tilapia* of the family Cichlidae represent the most important food fishes with potential for fish farming in tropical regions. They are very hardy, are easy to rear and can show good growth rates. In addition to all this they are attractive in market terms, having good flavour and a good flesh texture. Their importance as a source of protein in underdeveloped countries can hardly be overstated, particularly as in the main they are omnivorous, with distinct leanings towards vegetarianism.

The principal breeding strategies embrace two important facts, first, the well-known problem of over-reproduction and the need to avoid it, and, secondly, the availability of a wide range of species each with characteristic features of life styles (Wohlfarth and Hulata, 1983). Thus whereas salmonids offer a range of options based in part on distinctive geographic races, the tilapias are much more variable in terms of generic and specific attributes. In addition, the use of hybrids is commonplace but this also creates problems. Hybridization in hatcheries and in the wild can go beyond the F_1 stage and produce introgression. This 'mongrelization' is undesirable in that it is unregulated and will affect performance adversely, particularly where specific breeding programmes are undertaken to control sex ratio.

Sex ratio control to avoid the tendency of the tilapias to overbreed and generate stunted populations remains the principal goal of applied genetics in these species. The male-only techniques described in Chapter 11 and the genetic basis covered in Chapter 8 are currently applied in tilapias for population control but simple, reliable techniques for application in commercial farming at all levels remain to be devised. This development seems assured, but the regulation of tilapia broodstocks by the identification of species composition, together with the avoidance of introgressive hybridization of stock with inappropriate sex-determining mechanisms, are essential prerequisites of success.

The genetic diversity of tilapias is such that developments could rival the range of achievements in aquarium and pond species, but the primary goal must remain the regulation of population growth and clear systematic orientation of farm practice.

References

Agellon, L.B., Emery, C.J., Jones, J.M., Davies, S.L., Dingle, A.D. and Chen, T.T. (1988) Promotion of rapid growth of rainbow trout (*Salmo gairdneri*) by a recombinant fish growth hormone. *Canadian Journal of Fisheries and Aquatic Sciences*, **45**, 146–51.

Agnèse, J.F., Oberdorff, T. and Ozouf-Costaz, C. (1990) Karyotypic study of some species of family Mochokidae (Pisces, Siluriformes): evidence of female heterogamety. *Journal of Fish Biology*, **37**, 375–81.

Aida, T. (1921) On the inheritance of color in a freshwater fish, *Aplocheilus latipes*, Temminck and Schlegel, with special reference to sex linked inheritance. *Genetics, Princeton*, **6**, 554–73.

Aida, T. (1930) Further genetical studies of *Aplocheilus latipes*. *Genetics, Princeton*, **15**, 1–16.

Allendorf, F.W. (1975) Genetic variability in a species possessing extensive gene duplication: genetic interpretation of duplicate loci and examination of genetic variation in populations of rainbow trout. PhD thesis, University of Washington, Seattle, WA.

Alm, G. (1955) Artificial hybridization between different species of the salmon family. *Report of the Institute of Freshwater Research, Drottningholm*, **36**, 13–57.

Altukhov, Yu.P., Limansky, V.V., Payusova, A.N. and Truvella, K.A. (1969) Immunogenetic analysis of the intraspecific differentiation in the European anchovy (*Engraulis encrasicolus*) inhabiting the Black Sea and the Sea of Azov. I. Anchovy blood groups and a hypothetical mechanism of genetic regulation: heterogeneity of the Azov race. *Genetika (USSR)*, **5**, 50–64.

Ammerman, L.K. and Morizot, D.C. (1989) Biochemical genetics of endangered Colorado squawfish populations. *Transactions of the American Fisheries Society*, **118**, 435–40.

Anders, A., Anders, F. and Klinke, K. (1973) Regulation of gene expression in the Gordon-Kosswig melanoma system. I. The distribution of the controlling genes in the genome of the Xiphophorin fish, *Platypoecilus maculatus* and *Platypoecilus variatus*, in *Genetics and Mutagenesis of Fish* (ed J.H. Schroder), Springer-Verlag, Berlin, pp. 33–52.

Anders, F., Schartl, M., Barnekov, A. and Anders, A. (1984) *Xiphophorus* as an *in vivo* model for studies on normal and defective control of oncogenes. *Advances in Cancer Research*, **42**, 191–275.

Aquarist and Pondkeeper, **1-56** (1928-1991).

Arai, K. (1986) Effect of allotriploidization on development of the hybrids between female chum salmon and male brook trout. *Bulletin of the Japanese Society of Scientific Fisheries*, **52**, 823–30.

Arefjev, V.A. (1989) Karyotype variability in successive generations after hybridization between the great sturgeon *Huso huso* (L.), and the sterlet, *Acipenser ruthenus* L. *Journal of Fish Biology*, **35**, 819–28.

Atz, J.W. (1964) Intersexuality in fishes, in *Intersexuality in Vertebrates Including Man* (eds C.N. Armstrong and A.J. Marshall), Academic Press, London, pp. 145–232.

Avise, J.C. (1976) Genetics of plate morphology in an unusual population of threespine sticklebacks (*Gasterosteus aculeatus*). *Genetical Research*, **27**, 33–46.

Avtalion, R.R. and Don, J. (1990) Sex-determining genes in tilapia: a model of genetic recombination emerging from sex ratio results of three generations of diploid gynogenetic *Oreochromis aureus*. *Journal of Fish Biology*, **37**, 167–74.

Axelrod, H.R. and Vorderwinkler, W. (1957) *Encyclopedia of Tropical Fishes*, 2nd edn, T.F.H. Publications, Jersey City, NJ.

Ayles, G.B. and Baker, R.F. (1983) Genetic differences in growth and survival between strains and hybrids of rainbow trout (*Salmo gairdneri*) stocked in aquaculture lakes in the Canadian prairies. *Aquaculture*, **33**, 269–80.

Bailey, G.S., Wilson, A.C., Halver, J. and Johnson, C. (1970) Multiple forms of supernatant malate dehydrogenase in salmonid fishes: biochemical, immunological and genetic studies. *Journal of Biological Chemistry*, **245**, 5927–40.

Bakker, Th.C.M. and Sevenster, P. (1988) Plate morphs of *Gasterosteus aculeatus* Linnaeus (Pisces: Gasterosteidae): comments on terminology. *Copeia*, **1988**, 659–63.

Bakker, Th.C.M., Feuth-De Bruijn, E. and Sevenster, P. (1988) Albinism in the threespine stickleback, *Gasterosteus aculeatus*. *Copeia*, **1988**, 236–8.

Baldwin, L.A. and Busack, C.A. (1990) Induction of triploidy in white crappie by temperature shock, *Transactions of the American Fisheries Society*, **119**, 438–44.

Balsano, J.S., Darnell, R.M. and Abramoff, P. (1972) Electrophoretic evidence of triploidy associated with populations of the gynogenetic teleost *Poecilia formosa*. *Copeia*, **1972**, 292–7.

Banerjee, S.K., Misra, K.K., Banerjee, S. and Ray-Chaudhuri, S.P. (1988) Chromosome numbers, genome sizes, cell volumes and evolution of snake-head fish (family Channidae). *Journal of Fish Biology*, **33**, 281–9.

Barker, C.J. (1972) A method for the display of chromosomes of plaice *Pleuronectes platessa* and other marine fishes. *Copeia*, **1972**, 365–8.

Bartley, D.M., Gall, G.A.E. and Bentley, B. (1990) Biochemical genetic detection of natural and artificial hybridization of chinook and coho salmon in northern California. *Transactions of the American Fisheries Society*, **119**, 431–7.

Bartnick, V.G. (1970) Reproductive isolation between two sympatric dace, *Rhinichthys atrotulus* and *R. cataractae*, in Manitoba. *Journal of the Fisheries Research Board of Canada*, **27**, 2125–41.

Baynes, S.M. and Scott, A.P. (1982) Cryopreservation of rainbow trout spermatozoa: variation in membrane composition may influence spermatozoan survival, in *International Symposium on the Reproductive Physiology of Fish, Wageningen*, (eds C.J.J. Richter and H.J.T. Goos), Pudoc, Wageningen, Netherlands p. 128.

Baynes, S.M. and Scott, A.P. (1987) Cryopreservation of rainbow trout spermatozoa: the influence of sperm quality, egg quality and extender composition on post-thaw fertility. *Aquaculture*, **66**, 53–67.

Beamish, R.J., Merriless, M.J. and Crossman, E.J. (1971) Karyotypes and DNA values for members of the suborder *Esocoidei (Osteichthyes: Salmoniformes)*. *Chromosoma*, **34**, 436–47.

Beatty, R.A. (1964) Gynogenesis in vertebrates: fertilization by genetically inactivated spermatozoa, in *Effects of Ionizing Radiation on the Reproductive System*, Proceedings of an International Symposium, Fort Collins, Colorado, 1962 (eds. Carlson, W.D. and Gassner, F.X.). Pergamon Press, Oxford, pp. 229–38.

Beck, M.L., Biggers, C.J. and Dupree, H.K. (1983) Electrophoretic analysis of protein systems of *Ctenopharyngodon idella* (Val.), *Hypophthalmichthys nobilis* (Rich.) and their F_1 triploid hybrid. *Journal of Fish Biology*, **22**, 603–11.

Bell, M.A. (1984) Evolutionary phenetics and genetics. The threespine stickleback, *Gasterosteus aculeatus*, and related species, in *Evolutionary Genetics of Fishes* (ed. B.J. Turner), Plenum Press, New York and London, pp. 431–528.

Bellamy, A.W. (1922) Sex linked inheritance in the teleost *Platypoecilus maculatus* Günth. *Anatomical Record*, **24**, 419–20.

Bellamy, A.W. (1924) Bionomic studies on certain teleosts (Poeciliinae). I. Statement of problems, description of material, and general notes on life histories and breeding behaviour under laboratory conditions. *Genetics*, **Princeton, 9**, 513–29.

Bellamy, A.W. (1928) Bionomic studies on certain teleosts (Poeciliinae). II. Colour pattern inheritance and sex in *Platypoecilus maculatus*. *Genetics, Princeton*, **13**, 226–32.

Bellamy, A.W. and Queal, M.L. (1951) Heterosomal inheritance and sex determination in *Platypoecilus maculatus*. *Genetics*, **Princeton**, **36**, 93–107.

Bennington, N.L. (1936) Germ cell origin and spermatogenesis in the Siamese fighting fish, *Betta splendens*. *Journal of Morphology*, **60**, 103–25.

Bertin, L. (1925) Recherches bionomiques, biometriques et systematiques sur les epinoches (Gasterosteids). *Annales de l'Institut Oceanographique, Monaco*, **2**, 1–204.

Bertin, L. (1956) *Eels - a Biological Study*, Cleaver-Hume Press, London.

Bianco, P.G. (1982) Hybridization between *Alburnus albidus (C)* and *Leuciscus cephalus cabeda R.* in Italy. *Journal of Fish Biology*, **21**, 593–603.

Bianco, P.G. (1988) *Leuciscus cephalus* (Linnaeus), with records of fingerling adult males, *Leuciscus pleurobipunctatus* (Stephanidis) and their hybrids from western Greece. *Journal of Fish Biology*, **32**, 1–16.

Bidwell, C.A., Chrisman, C.L. and Libey, G.S. (1985) Polyploidy induced by heat shock in channel catfish. *Aquaculture*, **51**, 25–32.

Billard, R. (1977) Utilization d'un system tris-glycocolle pour tamponner le diluent d'insemination pour truite. *Bulletin Francais de Pisciculture* **264**, 102–12.

Billington, N., Danzmann, R.G., Hebert, P.D.N. and Ward, R.D. (1991) Phylogenetic relationships among four members of *Stizostedion* (Percidae) determined by mitochondrial DNA and allozyme analyses. *Journal of Fish Biology*, **39**, (Supplement A), 251–58.

Bishop, R.D. (1968) Evaluation of the striped bass (*Roccus saxatilis)* and white bass (*R. chrysops)* hybrids after two years. *Proceedings of the Annual Conference of the Southeastern Association of Game and Fish Commissioners*, **21**, 245–54.

Blacher, L.J. (1927) Materials for the genetics of *Lebistes reticulatus*. *Transactions of the Laboratory of Experimental Biology, Zoopark, Moscow*, **3**, 139–52.

Blacher, L.J. (1928) Materials for the genetics of *Lebistes reticulatus*. II. *Transactions of the Laboratory of Experimental Biology, Zoopark Moscow*, **4**, 244–52.

Blaxter, J.H.S. (1953) Sperm storage and cross-fertilization of spring and autumn spawning herring. *Nature, London*, **172**, 1189–90.

Bonar, S.A., Thomas, G.L. and Pauley, G.B. (1988) Evaluation of the separation of triploid and diploid grass carp, *Ctenopharyngodon idella* (Valenciennes) by external morphology. *Journal of Fish Biology*, **33**, 895–8.

Bondari, K. (1981) A study of abnormal characteristics of channel catfish and blue tilapia. *Proceedings of the Annual Conference of the Southeastern Association of Fish and Wildlife Agencies*, **35**, 568–80.

Bondari, K. (1983) Response to bidirectional selection for body weight in channel catfish. *Aquaculture*, **33**, 73–81.

Borowski, R. (1984) The evolutionary genetics of *Xiphophorus*, in *Evolutionary Genetics of Fishes* (ed. B.J. Turner). Plenum Press, New York, pp. 235–310.

Braem, R.A. and King, E.L. (1971) Albinism in lampreys in the upper Great Lakes. *Copeia*, **1971**, 176–9.

Brem, G., Brenig, B., Hörstgen-Schwark, G. and Winnaker, G.L. (1988) Gene transfer in tilapia *(Oreochromis niloticus)*. *Aquaculture*, **68**, 209–19.

Brieder, H. (1935) Uber Aussenfaktoren, die das Geslechtsverhaltnis bei *Xiphophorus helleri* Heckel. *Zeitschrift fur Wissenschaftliche Zoologie*, **146**, 383–416.

Brody, T., Kirsht, D., Parag, G., Wohlfarth, G., Hulata, G. and Moav, R. (1979) Biochemical genetic comparison of the Chinese and European races of the common carp. *Animal Blood Groups and Biochemical Genetics*, **10**, 141–9.

Bruslé, S. (1987) Sex-inversion of the hermaphroditic protogynous teleost *Coris julis* L. (Labridae). *Journal of Fish Biology*, **30**, 605–16.

Burrough, R.J. (1981) Problems of natural hybridization amongst cyprinids, in *Proceedings of the Second British Freshwater Fisheries Conference, University of Liverpool*. Janssen Services, Chislehurst, pp. 162–70.

Burtzev, I.A. (1972) Progeny of intergeneric hybrids of beluga and sterlet, in *Genetics Selection and Hybridization* (ed. Y. Sobel), Keter Press, Jerusalem, pp. 211–20.

Burtzev, I.A., Nikolaev, A.I. and Slizchenko, A.G. (1985) Russian × Siberian sturgeon hybrid (*Acipenser guldenstadti* Br. × *A. baeri* Br.), a new candidate for commercial culture, in *Culture of Sea Organisms* (ed. A.F. Karpevich), Collected Papers, VNIRO, 112–16.

Busack, C.A. (1983) Four generations of selection for high 56-day body weight in the mosquitofish (*Gambusia affinis*). *Aquaculture*, **33**, 83–7.

Buss, K. and Wright, J.E. (1958) Appearance and fertility of trout hybrids. *Transactions of the American Fisheries Society*, **87**, 172–81.

Bye, V.J. (1982) Methods for the production of female and sterile trout, in *Commercial Trout Farming* (ed. M.J. Bulleid), Janssen Services, Chislehurst, pp. 75–85.

Bye, V.J. (1984) The role of environmental factors in the timing of reproductive cycles, in *Fish Reproduction: Strategies and Tactics* (eds G.W. Potts and R.J. Wootton), Academic Press, London, pp. 187–205.

Bye, V.J. and Lincoln, R.F. (1979) Male trout, an expensive luxury. *Fish Farmer*, 2, 19–20.

Bye, V.J. and Lincoln, R.F. (1981) Get rid of the males and let the females prosper. *Fish Farmer*, **4**, 22–4.

Bye, V.J. and Lincoln, R.F. (1986) Commercial methods for the control of sexual maturation in rainbow trout (*Salmo gairdneri*). *Aquaculture*, **57**, 299–309.

Campos, H.H. (1972) Karyology of three galaxiid fishes, *Galaxias maculatus, G. platei* and *Brachigalaxias vullocki*. *Copeia*, **1972**, 368–70.

Campton, D.E. and Johnston, J.M. (1985) Electrophoretic evidence for a genetic admixture of native and nonnative rainbow trout in the Yakima River, Washington. *Transactions of the American Fisheries Society*, **114**, 782–93.

Campton, D.E., Allendorf, F.W., Behnke, R.J. and Utter, F.M. (1991) Comments: reproductive success of hatchery and wild steelhead. *Transactions of the American Fisheries Society*, **120**, 822–27.

Carbone, P., Vitturi, R., Catalano, E. and Macaluso, M. (1987) Chromosome sex determination and Y-autosome fusion in *Blennius tentacularis* Brunnich, 1765 (Pisces: Blennidae). *Journal of Fish Biology*, **31**, 597–602.

Carlson, D.M., Kettler, M.K., Fisher, S.E. and Whitt, G.S. (1982) Low genetic variability in paddlefish populations. *Copeia*, **1982**, 721–25.

Caspersson, T., Zech, L., Modest, E.J., Foley, G.E., Wagh, U. and Simonsson, E. (1969) DNA-binding fluorochromes for the study of the organisation of the metaphase nucleus. *Experimental Cell Research*, **58**, 141–52.

Chan, S.T.H. (1970) Natural sex reversal in vertebrates. *Philosophical Transactions of the Royal Society, London B*, **259**, 59–71.

Chan, S.T.H., Tang, F. and Lofts, B. (1972) The role of sex steroids on natural sex reversal in *Monopterus albus* (Pisces: Teleostei). Excerpta Medica, International Congress Series, **256**, 348.

Chao, N.H., Chen, H.P. and Liao, I.C. (1975) Study on cryogenic preservation of grey mullet sperm. *Aquaculture*, **5**, 389–406.

Chen, F.Y. (1969) Preliminary studies on the sex determining mechanism of *Tilapia mossambica* Peters and *T. hornorum* Trewavas. *Verhandlungen der Internationalen Vereingung für Theoretische und Angewandte Limnologie*, **17**, 719–24.

Chen, S.C. (1928) Transparency and mottling, a case of Mendelian inheritance in the goldfish. *Genetics, Princeton*, **13**, 434–52.

Chen T.R. (1969) Karyological heterogamety of deep sea fishes. *Postilla*, **130**, 1–29.

Chen, T.R., (1971) A comparative chromosome study of twenty killifish species of the genus *Fundulus* (Teleostei: Cyprinodontidae). *Chromosoma*, **32**, 436–53.

Chen, T.R. and Ebeling, A.W. (1968) Karyological evidence of female heterogamety in the mosquito fish, *Gambusia affinis*. *Copeia*, **1968**, 70–75.

Chen, T.R. and Ebeling, A.W. (1971) Chromosomes of the goby fishes in the genus *Gillichthys*. *Copeia*, **1971**, 171–4.

Chen, T.R. and Reisman, H.M. (1970) A comparative study of the North American species of stickleback (Teleostei, Gasterosteidae). *Cytogenetics*, **9**, 321–32.

Cherfas, N.B. (1966) Natural triploidy in the females of the unisexual variety of the goldfish *Carassius auratus gibelio* (Bloch). *Soviet Genetics*, **13**, 557–63.

Cherfas, N.B. (1977) Investigations on radiation-induced gynogenesis in the carp. II. Segregation with respect to several morphological characters in gynogenetic progenies. *Genetika (USSR)*, **13**, 811–20.

Cherfas, N.B. (1981) Gynogenesis in fishes, in *Genetic Basis of Fish Selection* (ed. V.S. Kirpichnikov), Springer-Verlag, Berlin, pp. 255–73.

Cherfas, N.B., Abramenko, M.I., Emel'yanova, O.V., Il'ina, I.D. and Truveller, K.A. (1985) Genetic features of induced gynogenesis in hybrids of the silver goldfish and carp. *Soviet Genetics*, **22**, 109–14.

Chevassus, B. and Dorson, M. (1990) Genetics of resistance to disease in fishes. *Aquaculture*, **85**, 83–107.

Chevassus, B., Guyomard, R., Chourrout, D. and Quillet, E. (1983) Production of viable hybrids in salmonids by triploidization. *Genetique Selection/Evolution*, **15**, 519–32.

Chiarelli, B., Ferrantelli, O. and Cuchi, C. (1969) The karyotype of some teleostean fish obtained by tissue culture in vitro. *Experientia*, **25**, 426–7.

Chilcote, M.W., Leider, S.A. and Lock, J.J. (1986) Differential reproductive success of hatchery and wild summer-run steelhead under natural conditions. *Transactions of the American Fisheries Society*, **115**, 726–35.

Chilcote, M.W., Leider, S.A. and Lock, J.J. (1991) Comments: reproductive success of hatchery and wild steelhead. *Transactions of the American Fisheries Society*, **120**, 816–22.

Chourrout, D. (1980) Thermal induction of diploid gynogenesis and triploidy in the eggs of the rainbow trout (*Salmo gairdneri* Richardson). *Reproduction Nutrition Development*, **20 (3A)**, 727–33.

Chourrout, D. (1982) Gynogenesis caused by ultraviolet irradiation of salmonid sperm. *Journal of Experimental Zoology*, **223**, 175–181.

Chourrout, D. (1984) Pressure-induced retention of second polar body and suppression of first cleavage in rainbow trout: production of all-triploids, all-tetraploids, and heterozygous and homozygous diploid gynogenetics. *Aquaculture*, **36**, 111–26.

Chourrout, D. and Happe, A. (1986) Improved methods of direct chromosome preparation in rainbow trout. *Aquaculture*, **52**, 255–61.

Chourrout, D.R., Guyomard, R. and Houdebine, L.M. (1986) High efficiency gene transfer in rainbow trout (*Salmo gairdneri* Rich.) by microinjection into egg cytoplasm. *Aquaculture*, **51**, 143–50.

Christie, W.J. (1960) Variation in vertebral count in F_2 hybrids of *Salvelinus fontinalis* × *S. namaycush*. *Canadian Journal of Fish Culture*, **26**, 15–21.

Cimino, M.C. (1972a) Egg production, polyploidization and evolution in a diploid all-female fish of the genus *Poeciliopsis*. *Evolution*, **26**, 294–306.

Cimino, M.C. (1972b) Meiosis in triploid all-female fish (*Poeciliopsis*: Poeciliidae). *Science, NY*, **175**, 1484–5.

Clemens, H.P. and Inslee, T. (1968) The production of unisexual broods by *Tilapia mossambica* sex-reversed with methyltestosterone. *Transaction of the American Fisheries Society*, **97**, 18–21.

Cognie, F., Billard, R. and Chao, N.H. (1989) La cryoconservation de la laitance de la carpe, *Cyprinus carpio*. *Journal of Applied Ichthyology*, **5**, 165–76.

Collares-Pereira, M.J. (1985) The "*Rutilus alburnoides* (Steindacher, 1966) complex" (Pisces, Cyprinidae). II. First data on the karyology of a well established diploid-triploid group. *Arquivos Museu Bocage*, **A 3**, 69–90.

Collares-Pereira, M.J. and Coelho, M.M. (1983) Biometrical analysis of *Chondrostoma polylepis* × *Rutilus arcasi* natural hybrids (Osteichthyes-Cypriniformes-Cyprinidae) *Journal of Fish Biology*, **23**, 495–509.

Comings, D.E. (1978) Mechanisms of chromosome banding and implications for chromosome structure. *Annual Review of Genetics*, **12**, 25–46.

Cousin-Gerber, M., Burger, G., Boisseau, C. and Chevassus, B. (1989) Effect of methyltestosterone on sex differentiation and gonad morphogenesis in rainbow trout *Oncorhynchus mykiss*. *Aquatic living Resources*, **2**, 225–30.

Crivelli, A.J. and Dupont, F. (1987) Biometrical and biological features of *Alburnus alburnus × Rutilus rubilio* natural hybrids from Lake Mikra Prespa, Northern Greece. *Journal of Fish Biology*, **31**, 721–33.

Cuellar, O. and Uyeno, T. (1972) Triploidy in rainbow trout. *Cytogenetics*, **11**, 508–15.

Cumming, K.B. (1967) Natural hemagglutinins in marine fishes. *Research Bulletin of the International Commission for the North West Atlantic Fisheries*, **4**, 59–66.

Cushing, D.H. (1988) *The Provident Sea*, Cambridge University Press, Cambridge.

Cushing, J.E. and Durall, G.L. (1957) Examination of the blood of *Ictalurus n. nebulosus* Le Seur, *I. p. punctatus* Le Seur, *Geneonemus lineatus* Ayres, and *Katsuwonus pelamis* Linnaeus. *American Naturalist*, **91**, 121–6.

Dangel, J.R. (compiler) (1973) An annotated bibliography of interspecific hybridization of Salmonidae. *F.A.O Fisheries Circulars*, **133**, 1–32.

Danzmann, R.G., Van der Kraak, G.J., Chen, T.T. and Powers, D.A. (1990) Metabolic effects of bovine growth hormone and genetically engineered rainbow trout growth hormone in rainbow trout *(Oncorhynchus mykiss)* reared at a high temperature. *Canadian Journal of Fisheries and Aquatic Sciences*, **47**, 1292–1301.

Darlington, C.D. and La Cour, L.F. (1950) *The Handling of Chromosomes*, 2nd edn, George Allen and Unwin, London.

Davidson, W.S., Birt, T.P. and Green, J.M. (1989) A review of genetic variation in Atlantic salmon, *Salmo salar* L., and its importance for stock identification, enhancement programmes and aquaculture. *Journal of Fish Biology*, **34**, 547–60.

Davis, L.G., Dibner, M.D. and Battey, J.E. (1986) *Basic Methods in Molecular Biology*, Elsevier, New York.

Dawley, R.M., Schultz, R.J. and Goddard, K.A. (1987) Clonal reproduction and polyploidy in unisexual hybrids of *Phoxinus eos* and *Phoxinus neogaeus* (Pisces: Cyprinidae). *Copeia*, **1987**, 275–83.

DeLigny, W. (1969) Serological and biochemical studies on fish populations. *Oceanography and Marine Biology, Annual Review*, **7**, 411–513.

De Vlaming, V.L. (1972) Environmental control of teleost reproductive cycles: a brief review. *Journal of Fish Biology*, **15**, 71–91.

Dhar, N.J. and Chatterjee, K. (1984) Chromosomal evolution in Indian murrels *(Channiformes: Channidae)*. *Caryologia*, **37**, 359–71.

Dickerson, R.E. and Geis, I. (1983) *Haemoglobin: Structure, Function, Evolution and Pathology*. The Benjamin/Cummings Publishing Co., Menlo Park, CA.

Donaldson, E.M. and Hunter, G.A. (1983) Induced final maturation, ovulation and spermiation in cultured fish, in *Fish Physiology*, Vol. 9B (eds W.S. Hoar, D.J. Randall and E.M. Donaldson), Academic Press, New York, pp. 351–403.

Donaldson, L.R. and Olson, P.R. (1957) Development of rainbow trout broodstock by selective breeding. *Transactions of the American Fisheries Society*, **85**, 93–101.

Down, N.E., Donaldson, E.M., Dye, H.M., Boone, T.C., Langley, K.E. and Souza, L.M. (1989) A potent analogue of recombinant bovine somatotropin accelerates growth in juvenile coho salmon *(Oncorhynchus kisutch)*. *Canadian Journal of Fisheries and Aquatic Sciences*, **46**, 178–83.

Duchac, B.J. and Buhler, E.M. (1983) The expression of H-Y antigen in the sex-change fish *Coris julis*. *Experientia*, **39**, 767–9.

Dunham, R.A. and Smitherman, R.O. (1983) Response to selection and realised heritability for body weight in three strains of channel catfish, *Ictalurus punctatus*, grown in earthen ponds. *Aquaculture*, **33**, 89–96.

Dunham, R.A., Eash, J., Askins, J. and Townes, T.M. (1987) Transfer of metallothionein-human growth hormone fusion gene into channel catfish. *Transactions of the American Fisheries Society*, **116**, 87–91.

Ebeling, A.W. and Chen, T.R. (1970) Heterogamety in teleostean fishes. *Transactions of the American Fisheries Society*, **99**, 131–8.

Ebeling, A.W. and Seltzer, P.Y. (1971) Cytological confirmation of female homogamety in the deep-sea fish *Bathylagus milleri*. *Copeia*, **1971**, 560–62.

Echelle, A.A. and Mosier, D.T. (1981) All-female fish: a cryptic species of *Menidia* (Atherinidae). *Science, NY*, **212**, 1411–13.

Echelle, A.A., Echelle, A.F., DeBault, L.E. and Durham, D.W. (1988) Ploidy levels in silverside fishes (Atherinidae, *Menidia*) on the Texas coast: flow-cytometric analysis of the occurrence of allotriploidy. *Journal of Fish Biology*, **32**, 835–44.

Economidis, P.S. and Sinis, A.I. (1988) A natural hybrid of *Leuciscus cephalus macedonicus* × *Chalcalburnus chalcoides macedonicus* (Pisces, Cyprinidae) from Lake Volvi (Macedonia, Greece). *Journal of Fish Biology*, **32**, 593–605.

Economidis, P.S. and Wheeler, A. (1989) Hybrids of *Abramis brama* with *Scardinius erythrophthalmus* and *Rutilus rutilus* from Lake Volvi, Macedonia, Greece. *Journal of Fish Biology*, **35**, 295–9.

Edds, D.R. and Echelle, A.A. (1989) Genetic comparisons of hatchery and natural stocks of small endangered fishes: Leon Springs pupfish, Comanche Springs pupfish, and *Pecos gambusia*. *Transactions of the American Fisheries Society*, **118**, 441–6.

Ehlinger, N.F. (1964) Selective breeding of trout for resistance to furunculosis. *New York Fish and Game Journal*, **11**, 78–90.

Ehlinger, N.F. (1977) Selective breeding of trout for resistance to furunculosis. *New York Fish and Game Journal*, **24**, 25–36.

Evans, H.J. (1965) A simple microtechnique for obtaining human chromosome preparations with some comments on DNA replication in sex-chromosomes of the goat, cow and pig. *Experimental Cell Research*, **38**, 511–16.

Falconer, D.S. (1981) *An Introduction to Quantitative Genetics*, 2nd edn, Oliver and Boyd, Edinburgh.

Fan, Z. and Fox, D.P. (1991) Robertsonian polymorphism in plaice, *Pleuronectes platessa* L., and cod, *Gadus morhua* L., (Pisces Pleuronectiformes and Gadiformes). *Journal of Fish Biology*, **38**, 635–40.

Fankhauser, G. (1945) The effects of changes in chromosome number on amphibian development. *Quarterly Review of Biology*, **20**, 20–78.

FAO (1990) Yearbook of Fishery Statistics, **66**, 1988, 1–502.

Feddern, H.A. (1968) Hybridization between the western Atlantic angelfishes *Holacanthus isabelita* and *H. ciliaris*. *Bulletin of Marine Science of the University of Hawaii*, **18**, 351–82.

Ferguson, A. (1980) *Biochemical Systematics and Evolution*, Blackie, Glasgow and London.

Ferguson, A. and Fleming, C.C. (1983) Evolutionary and taxonomic significance of protein variation in the brown trout (*Salmo trutta* L.) and other salmonid fishes, in *Protein Polymorphism: Adaptive and Taxonomic Significance* (eds G.S. Oxford and D. Rollinson), Academic Press, London, pp. 85–99.

Ferguson, A. and Mason, F.M. (1981) Allozyme evidence for reproductively isolated sympatric populations of brown trout *Salmo trutta* L. in Lough Melvin, Ireland. *Journal of Fish Biology*, **18**, 629–42.

Ferris, S.D. (1984) Tetraploidy and the evolution of the catostomid fishes, in *Evolutionary Genetics of Fishes* (ed. B.J. Turner), Plenum Press, New York, pp. 55–93.

Ferris, S.D. (1987) The utility of mitochondrial DNA in fish genetics and fishery management, in *Population Genetics and Fishery Management* (ed. N. Ryman and F.M. Utter) Seattle, University of Washington Press, 277–99.

Fishelson, L. (1970) Protogynous sex reversal in *Anthias squamipinnis* (Teleostei, Anthiidae) regulated by the presence or absence of a male fish. *Nature, Lond.*, **227**, 90–91.

Fevolden, S.E. and Haug, T. (1988) Genetic population structure of Atlantic halibut, *Hippoglossus hippoglossus*. *Canadian Journal of Fisheries and Aquatic Sciences*, **45**, 2–7.

Fletcher, G.L., Shears, M.A., King, M.J., Davies, P.L. and Choy, L.H. (1988) Evidence for antifreeze protein gene transfer in Atlantic salmon (*Salmo salar*). *Canadian Journal of Fisheries and Aquatic Sciences*, **45**, 352–7.

Foerster, R.E. (1968) The sockeye salmon. *Bulletin of the Fisheries Research Board of Canada* **162**, Ottawa.

Foerster, W. (1987) A new Guppy mutation. *Tropical Fish Hobbyist*, **36**,(6), 46–9.

Foley, J.O. (1926) The spermatogenesis of *Umbra limia* with special reference to the behaviour of the spermatogonial chromosomes and the first maturation division. *Biological Bulletin*, **50**, 117–47.

Formacion, M.J. and Uwa, H. (1985) Cytogenetic studies on the origin and species differentiation of the Philippine medaka, *Oryzias luzonensis*. *Journal of Fish Biology*, **27**, 285–91.

Fraser, A.C. and Gordon, M. (1928) Crossing-over between the W and Z chromosomes of the killifish *Platypoecilus*. *Science, NY*, **67**, 470.

Fraser, A.C. and Gordon, M. (1929) The genetics of *Platypoecilus*. The linkage of two sex-linked characters. *Genetics*, **14**, 160–179.

Friars, G.W., Bailey, J.K. and Coombs, K.A. (1990) Correlated responses to selection for grilse length in Atlantic salmon. *Aquaculture*, **85**, 171–6.

Friedman, B. and Gordon, M. (1934) Chromosome number in Xiphophorin fishes. *American Naturalist*, **6**, 446–55.

Fries, L.T. and Harvey, W.D. (1989) Natural hybridization of white bass and yellow bass in Texas. *Transactions of the American Fisheries Society*, **118**, 87–9.

Fujino, K. and Kang, T. (1968) Transferrin groups of tunas. *Genetics*, **59**, 79–91.

Garcia de Leaniz, C. and Verspoor, E. (1989) Natural hybridization between Atlantic salmon, *Salmo salar*, and brown trout, *Salmo trutta*, in northern Spain. *Journal of Fish Biology*, **33**, 41–6.

Garcia-Vazquez, E., Linde, A.R., Blanco, G., Sanchez, J.A., Vazquez, E. and Rubio, J. (1988) Chromosome polymorphism in farm fry stocks of Atlantic salmon from Asturias. *Journal of Fish Biology*, **33**, 581–7.

GESAMP (1990) IMO/FAO/UNESCO/WMO/WHO/IAEA/UN/UNEP Joint Group of Experts on the Scientific Aspects of Marine Pollution: The state of the marine environment. *Reports and Studies of GESAMP*, **39**, 1–111.

Ginsburg, A.S. (1968) *Fertilization in Fishes and the Problem of Polyspermy*, Israel Program for Scientific Translations, Jerusalem.

Gjedrem, T. (1979) Selection for growth rate and domestication in Atlantic salmon. *Zeitschrift fur Tierzuchtung und Zuchtungsbiologie*, **96**, 56–9.

Goetz, F.W., Donaldson, E.M., Hunter, G.A. and Dye, H.M. (1979) Effects of estradiol-17β and 17α-methyltestosterone on gonadal differentiation in the coho salmon, *Oncorhynchus kisutch*. *Aquaculture*, **17**, 267–78.

Gold, J.R., Li, Y.C., Shipley, N.S. and Powers, P.K. (1990) Improved methods for working with fish chromosomes with a review of metaphase chromosome banding. *Journal of Fish Biology*, **37**, 563–75.

Golovinskaya, K.A. (1968) Genetics and selection of fish and artificial gynogenesis of the carp (*Cyprinus carpio*). *F.A.O Fisheries Reports*, (**44**) **4**, 215–22.

Gomel'skii, B.I., Cherfas, N.B and Emel'yanova, O.V. (1985) Ability of *Carassius–Cyprinus* hybrids to produce diploid spermatozoa. *Doklady Biological Sciences*, **282**, 357–60.

Goodpasture, C. and Bloom, S.E. (1975) Visualization of nucleolar organiser regions in mammalian chromosomes using silver staining. *Chromosoma*, **53**, 37–50.

Goodrich, H.B. and Smith, M.A. (1937) Genetics and histology of the color pattern in the normal and albino paradise fish, *Macropodus opercularis* L. *Biological Bulletin*, **73**, 527–34.

Goodrich, H.B., Harrison, R.W., Josephson, N.D. and Trinkhaus, J.P. (1943) Three genes in *Lebistes reticulatus*. *Genetics, Princeton*, **28**, 75.

Goodrich, H.B., Hill, G.A. and Arrick, M.S. (1941) The chemical identification of gene-controlled pigments in *Platypoecilus* and *Xiphophorus* and comparison with other tropical fish. *Genetics, Princeton,* **26**, 573–86.

Goodrich, H.B., Josephson, N.D., Trinkaus, J.P. and Slate, J.M. (1944) The cellular expression and genetics of two new genes in *Lebistes reticulatus. Genetics, Princeton,* **29**, 584–92.

Gordon, J.W., Scangos, G.A., Plotkin, D.J., Barbara, J.A. and Rudle, F.H. (1980) Genetic transformation of mouse embryos by microinjection of purified DNA. *Proceedings of the National Academy of Sciences, U.S.A.,* **77**, 7380–84.

Gordon, M. (1927) The genetics of a viviparous top-minnow *Platypoecilus;* the inheritance of two kinds of melanophores. *Genetics, Princeton,* **12**, 253–83.

Gordon, M. (1931) Hereditary basis of melanosis in hybrid fishes. *American Journal of Cancer,* **15**, 1495–1523.

Gordon, M. (1937) Genetics of *Platypoecilus.* III. Inheritance of sex and crossing-over of the sex chromosomes in the platyfish. *Genetics,* **Princeton, 22**, 376–92.

Gordon, M. (1942) Mortality of albino embryos and aberrant Mendelian ratios in certain broods of *Xiphophorus helleri. Zoologica,* **27**, 73–4.

Gordon, M. (1951) A loan repaid with interest. *Animal Kingdom,* **54**, 173–4.

Graves, J.E., Ferris, S.D. and Dizon, A.E. (1984) Close genetic similarity of Atlantic and Pacific skipjack tuna *(Katsuwonus pelamis)* demonstrated with restriction endonuclease analysis of mitochondrial DNA. *Marine Biology,* **79**, 315–91.

Grobstein, C. (1948) Optimum gonopodial morphogenesis in *Platypoecilus maculatus* with constant dosage of methyltestosterone. *Journal of Experimental Zoology,* **109**, 215–37.

Guerrero, R.D. (1975) Use of androgens for the production of all-male *Tilapia aurea* (Steindachner). *Transactions of the American Fisheries Society,* **104**, 342–48.

Gyllensten, U. and Wilson, A.C. (1987) Mitochondrial DNA of salmonids: inter- and intraspecific variability detected with restriction enzymes, in *Population Genetics and Fishery Management* (eds N. Ryman and F. Utter), University of Washington Press, Seattle, WA, pp. 301–17.

Haaf, T. and Schmid, M. (1984) An early stage of ZW/ZZ sex chromosome differentiation in *Poecilia sphenops* var. *melanistica* (Poeciliidae, Cyprinodontiformes). *Chromosoma,* **89**, 37–41.

Hagen, D.W. and Gilbertson, L.G. (1972) Geographic variations and environmental selection in *Gasterosteus aculeatus* L. in the Pacific Northwest, America. *Evolution,* **26**, 32–51.

Hagen, D.W. and Gilbertson, L.G. (1973) The genetics of plate morphs in freshwater threespine sticklebacks. *Heredity, Lond.,* **31**, 75–84.

Haldane, J.B.S. (1922) Sex ratio and unisexual sterility in hybrid animals. *Journal of Genetics,* **12**, 101–9.

Hallerman, E.M., Schneider, J.F., Gross, M.L., Faras, A.J., Hackett, P.B., Guise, K.S. and Kapuscinski, A.R. (1988) Enzymatic dechorionation of goldfish, walleye and northern pike eggs. *Transactions of the American Fisheries Society,* **117**, 456–60.

Hammerman, I.S. and Avtalion, R.R. (1979) Sex determination in *Sarotherodon* (Tilapia). Part 2: The sex ratio as a tool for the determination of genotype – a model of autosomal and gonosomal influence. *Theroretical and Applied Genetics,* **55**, 177–87.

Harrington, R.W.,Jun. (1961) Oviparous hermaphroditic fish with internal fertilization. *Science, NY,* **135**, 1749–50.

Harrington, R.W.,Jun. (1971) How ecological and genetic factors interact to determine whether self-fertilizing hermaphrodites of *Rivulus marmoratus* change into functional secondary males with a reappraisal of the modes of intersexuality among fishes. *Copeia,* **1971**, 389–432.

Harrington, R.W.,Jun. (1975) Sex determination and differentiation among the uni-parental homozygotes of the hermaphroditic fish *Rivulus marmoratus* (Cyprinodontidae, Atheriniformes), in *Intersexuality in the Animal Kingdom*, (ed. R. Reinboth), Springer-Verlag, Berlin, pp. 249–62.

Hartley, S.E. (1987) The chromosomes of salmonid fishes. *Biological Reviews*, **62**, 197–214.

Hartley, S.E. (1991) C, Q, and restriction enzyme banding of the chromosomes in brook trout (*Salvelinus fontinalis*) and Arctic charr (*Salvelinus alpinus*). *Hereditas*, **114**, 253–61.

Hartley, S.E. and Horne, M.T. (1984a) Chromosome polymorphism and constitutive heterochromatin in Atlantic salmon *Salmo salar. Chromosoma*, **89**, 377–80.

Hartley, S.E. and Horne, M.T. (1984b) Chromosome relationships in the genus *Salmo. Chromosoma*, **90**, 229–37.

Hartley, S.E. and Horne, M.T. (1985) Cytogenetic techniques in fish genetics. *Journal of Fish Biology*, **26**, 575–82.

Haskins, C.P. and Haskins, E.F. (1948) Albinism, a semilethal autosomal mutation in *Lebistes reticulatus. Heredity, Lond.*, **2**, 251–62.

Haskins, C.P., Haskins, E.F. and Hewitt, R.E. (1960) Pseudogamy as an evolutionary factor in the poeciliid fish *Mollienesia formosa. Evolution*, **14**, 473–83.

Haskins, C.P., Haskins, E.F., McLaughlin, J.J. and Hewitt, R.E. (1961) Polymorphism and population structure in *Lebistes reticulatus*, an ecological study. In *Vertebrate Speciation*, University of Texas Press, New York, pp. 320–95.

Haskins, C.P., Young, P., Hewitt, R.E. and Haskins, E.F. (1970) Stabilized heterozygosis of supergenes mediating certain Y-linked colour patterns in populations of *Lebistes reticulatus. Heredity, Lond.*, **25**, 575–89.

Haussler, G. (1928) Über die Melanombildung bei Barstarden von *Xiphophorus helleri* und *Platypoecilus maculatus* var *rubra. Klinische Wochenschrift*, **7**, 1561–2.

Hershberger, W.K., Myers, J.M., Iwamota, R.N. and McAuley, W.C. (1990a) Assessment of inbreeding and its implications for salmon broodstock development. *National Oceanic and Atmospheric Administration, of the National Marine Fisheries Service, Washington*, **92**, 1–7.

Hershberger, W.K., Myers, J.M., Iwamoto, R.N., McAuley, W.C. and Saxton, A.M. (1990b) Genetic changes in the growth of coho salmon (*Oncorhynchus kisutch*) in marine net-pens produced by ten years of selection. *Aquaculture*, **85**, 187–97.

Hertwig, O. (1911) Die radiumkrankheit tierischer Kiemzellen. *Archiv für Mikroscopische Anatomie und Entwicklungsmechanik*, **77**, 1–97.

Heuts, M.J. (1947) Experimental studies on adaptive evolution in *Gasterosteus aculeatus* L. *Evolution*, **1**, 89–102.

Hickling, C.F. (1960) The Malacca Tilapia hybrids. *Journal of Genetics*, **57**, 1–10.

Hines, N.O. (1976) *Fish of Rare Breeding: Salmon and Trout of the Donaldson Strains*, Smithsonian Institution Press, Washington, DC.

Hoar, W.S., Randall, D.J. and Donaldson E.M. (eds) (1983) *Fish Physiology*, Vols 9A and B, Academic Press, New York.

Hopkins, K.D., Shelton, W.L. and Engle, C.R. (1979) Estrogen sex reversal of *Tilapia aurea. Aquaculture*, **18**, 263–68.

Hörstgen-Schwark, G., Fricke, H. and Langholz, H.-J. (1986) The effect of strain crossing on the production performance in rainbow trout. *Aquaculture*, **57**, 141–52.

Huang, C.-M. and Liao, I.-C. (1990) Response to mass selection for growth rate in *Oreochromis niloticus. Aquaculture*, **85**, 199–205.

Hubbs, C.L. (1955) Hybridization between fish species in nature. *Systematic Zoology*, **4**, 1–20.

Hubbs, C.L. and Hubbs, L.C. (1932) Apparent parthenogenesis in nature, in a form of fish of hybrid orgin. *Science, NY*, **76**, 628–30.

Hubbs, C.L. and Kuronoma, K. (1942) Hybridization in nature between two genera of flounders in Japan. *Papers from the Michigan Academy of Science, Arts and Letters*, **27**, 267–306.

Hulata, G., Wohlfarth, G.W. and Halevy, A. (1986) Mass selection for growth rate in Nile tilapia (*Oreochromis niloticus*). *Aquaculture*, **57**, 177–84.

Hulbert, P.J. (1985) Post stocking performance of hatchery-reared yearling brown trout. *New York Fish and Game Journal*, **32**, 1–8.

Humphries, J.M., Bookstein, F.L., Chernoff, B., Smith, R., Elder, R.L. and Poss, S.G. (1981) Multivariant discrimination by shape in relation to size. *Systematic Zoology*, **30**, 291–308.

Hunter, G.A. and Donaldson, E.M. (1983) Hormonal sex control and its application to fish culture, in *Fish Physiology* Vol. 9B (eds W.S. Hoar, D.J. Randall and E.M. Donaldson), Academic Press, New York, pp. 223–303.

Hunter, G.A., Donaldson, E.M., Goertz, F.W. and Edgell, P.R. (1982) Production of all-female and sterile coho salmon and experimental evidence for male heterogamety. *Transactions of the American Fisheries Society*, **3**, 367–72.

Hunter, G.A., Donaldson, E.M., Stoss, J. and Baker, J. (1983) Production of monosex female groups of chinook salmon (*Oncorhynchus tshawytscha*) by the fertilization of normal ova with sperm from sex-reversed females. *Aquaculture*, **33**, 335–64.

Izyumov, Yu.G. and Gerasimenko, O.G. (1987) Experimental evidence of the hybrid origin of the Middle Volga population of bream, *Abramis brama*. *Journal of Ichthyology*, **27**, 101–3.

Jahn, L., Douglas, D.R., Terhaar, M.J. and Kruse, G.W. (1987) Effects of stocking hybrid striped bass in Spring Lake, Illinois. *North American Journal of Fisheries Management*, **7**, 522–30.

Jalabert, B., Billard, R. and Chevassus, B. (1975) Preliminary experiments on sex control in trout: production of sterile fishes and simultaneous self-fertilizable hermaphrodites. *Annales de Biologie Animale, Biochimie et Biophysique*, **15**, 19–28.

Jalabert, B., Moreau, J., Planquette, P. and Billard, R. (1974) Determinisme du sexe chez *Tilapia macrochir* et *Tilapia nilotica*: action de la methyltestosterone dans l'alimentation des alevins sur la differentiation sexuelle; proportion des sexes dans la descendance des males "inverses". *Annales de Biologie Animale, Biochimie et Biophysique*, **14**, 729–39.

Johnson, K.R., Wright, J.E. and May, B. (1987) Linkage relationships reflecting ancestral tetraploidy in salmonid fish. *Genetics*, **116**, 579–91.

Johnstone, R. (1985) Induction of triploidy in Atlantic salmon by heat shock. *Aquaculture*, **49**, 133–9.

Johnstone, R. and Lincoln, R.F. (1986) Ploidy estimation using erythrocytes from formalin-fixed salmonid fry. *Aquaculture*, **55**, 145–8.

Johnstone, R., Simpson, T.H. and Youngson, A.F. (1978) Sex reversal in salmonid culture. *Aquaculture*, **13**, 115–34.

Joyner, T. (1976) Farming ocean ranges for salmon. *Journal of the Fisheries Research Board of Canada*, **33**, 902–4.

Kajishima, T. (1977) Genetic and developmental analysis of some new color mutants in the goldfish, *Carassius auratus*. *Genetics*, **86**, 161–74.

Kallman, K.D. (1962a) Gynogenesis in the teleost *Mollienesia formosa* Girard, with a discussion on the detection of parthenogenesis in vertebrates by tissue transplantation. *Journal of Genetics*, **58**, 7–24.

Kallman, K.D. (1962b) Population genetics of the gynogenetic teleost, *Mollienesia formosa*. *Evolution*, **16**, 497–504.

Kallman, K.D. (1964) An estimate of the number of histocompatibility loci in the teleost *Xiphophorus maculatus*. *Genetics*, **50**, 583–95.

Kallman, K.D. (1965) Genetics and geography of sex determination in the poeciliid fish, *Xiphophorus maculatus*. *Zoologica*, **50**, 151–90.

Kallman, K.D. (1973) The sex-determining mechanism of the platyfish, *Xiphophorus maculatus*, in *Genetics and Mutagenesis of Fish* (ed. J.H. Schroder), Springer-Verlag, Berlin, pp.19–28.

Kallman, K.D. (1975) The platy fish, *Xiphophorus maculatus*, in *Handbook of Genetics*, vol.IV. Plenum Press, New York, pp 81–132.

Kallman, K.D. (1984) A new look at sex determination in poeciliid fishes, in *Evolutionary Genetics of Fishes* (ed. B.J. Turner), Plenum Press, New York, pp. 95–171.

Kallman, K.D. and Atz, J.W. (1966) Gene and chromosome homology in the fishes of the genus *Xiphophorus. Zoologica*, **51**, 107–35.

Kallman, K.D. and Harrington, R.W.,jnr. (1964) Evidence for the existence of homozygous clones in the self-fertilizing hermaphroditic teleost *Rivulus marmoratus* (Poey). *Biological Bulletin*, **126**, 101–14.

Katasonov, V.Ya., (1973) Investigation of color in hybrids of common and ornamental (Japanese) carp. I. Transmission of dominant color types. *Genetika (Moscow)* **9** (8), 59–69. (*Soviet Genetics* **9** (8), cover-to-cover translation).

Katasonov, V.Ya., (1974) II Pleiotropic effect of dominant color genes. *Genetika (Moscow)* **10** (12), 56–66 (*Soviet Genetics* **10** (12), cover-to-cover translation).

Katasonov, V.Ya., (1978) III Inheritance of blue and orange color types. *Genetika (Moscow)*, **14** (12), 2184–92 (*Soviet Genetics* **14** (12), cover-to-cover translation).

Kerby, J.H., Hinshaw, J.M. and Huish, M.T. (1987) Increased growth and production of striped bass × white bass hybrids in earthen ponds. *Journal of the World Aquaculture Society*, **18**, 35–43.

Kincaid, H.L. (1983) Inbreeding in fish populations used for acquaculture. *Aquaculture*, **33**, 215–27.

Kincaid, H.L., Bridges, W.R., and Limbach, B. von (1977) Three generations of selection for growth rate in fall-spawning rainbow trout. *Transactions of the American Fisheries Society*, **106**, 621–8.

King, D.P.F., Hovey, S.J., Thompson, D. and Scott, A. (1992) Mitochondrial DNA variation in Atlantic salmon (*Salmo salar* L.) populations. *Journal of Fish Biology* (in press).

Kinghorn, B.P. (1983) A review of quantitative genetics in fish breeding. *Aquaculture*, **31**, 283–305.

Kirpichnikov, V.S. (1981) *Genetic Basis of Fish Selection.* Springer-Verlag, Berlin.

Kirpichnikov, V.S. (1992) Adaptive nature of intrapopulational biochemical polymorphism in fish. *Journal of Fish Biology*, **40**, 1–16.

Klar, G.T., Parker, N.C. and Goudie, C.A. (1988) Comparison of growth among families of channel catfish. *The Progressive Fish-Culturist*, **50**, 173–8.

Kligerman, A.D. and Bloom, S.E. (1977) Rapid chromosome preparations from solid tissues of fishes. *Journal of the Fisheries Research Board of Canada*, **34**, 266–9.

Kobayashi, H. (1971) A cytological study on gynogenesis of the triploid ginbuna (*C. auratus langsdorfi*). *Japanese Journal of Ichthyology*, **22**, 234–40.

Koehn, R.K. (1970) Functional and evolutionary dynamics of polymorphic esterases in catostomid fishes. *Transactions of the American Fisheries Society*, **99**, 219–28.

Komen, J. and Richter, C.J.J. (1992) Sex control in common carp (*Cyprinus carpio* L.). In *Recent Advances in Aquaculture*, **4**.

Komen, J., Bongers, A.B.J., Richter, C.J.J., Muiswinkel, W.B. van and Huisman, E.A. (1991) Gynogenesis in common carp (*Cyprinus carpio* L). II: The production of homozygous gynogenetic clones and F1 hybrids. *Aquaculture*, **92**, 127–42.

Komen, J., Lodder, P.A.J., Huskens, F., Richter, C.J.J. and Huisman, E.A. (1989) Effects of oral administration of 17α-methyltestosterone and 17β-estradiol on gonadal development in common carp, *Cyprinus carpio* L. *Aquaculture*, **78**, 349–63.

Kosswig, C. (1929) Zur Frage der Geschwulstbildung bei Gattungsbastarden der Zohnkarpfen *Xiphophorus* und *Platypoecilus. Zeitschrift fur Induktive Abstammungs-Vererbungslehre*, **52**, 114–20.

Kosswig, C. (1935) Uber Albinismus bei Fischen. *Zoologischer Anzeiger*, **110**, 41–7.

Kosswig, C. (1964) Polygenic sex determination. *Experientia*, **20**, 190–99.

Kozlov, V.I. (1970) A natural hybrid of the spiny sturgeon [*A. rudiventris derjavini* (Borzenko)] and the Kura sevryuga [*A. stellatus* (Pollas)]. *Journal of Ichthyology*, **10**, 466–70.

Kramer, C.R., Koulish, S. and Bertacchi, P.L. (1988) The effects of testosterone implants on ovarian morphology in the bluehead wrasse, *Thalassoma bifasciatum* (Bloch) (Teleostei: Labridae). *Journal of Fish Biology*, **32**, 397–407.

Larson, G.L. and Moore, S.E. (1985) Encroachment of exotic rainbow trout into stream populations of native brook trout in the Southern Appalachian Mountains. *Transactions of the American Fisheries Society*, **114**, 195–203.

Law, R. (1991) Fishing in evolutionary waters. *New Scientist*, **129** (1758), 35–7.

Lerner, I.M. (1954) *Genetic Homeostasis*, Oliver and Boyd, Edinburgh.

Lessent, P. (1968) Essais d'hybridization dans le genre *Tilapia* à la station de recherches piscicole de Bouaké, Côte d'Ivoire. *FAO Fisheries Reports* (**44**) **4**, 148–59.

Leung, L.K.-P. and Jamieson, B.G.M. (1991) Live preservation of fish gametes in *Fish Evolution and Systematics: Evidence from Spermatozoa*. Cambridge University Press, Cambridge, pp. 245–269.

Liem, K.F. (1968) Geographical and taxonomic variation in the pattern of natural sex-reversal in the teleost fish Order Synbranchiformes. *Journal of Zoology*, **156**, 225–38.

Liley, N.R. (1966) Ethological isolating mechanisms in four sympatric species of poeciliid fishes. *Behaviour*, (Supp). **13**, 1–97.

Lincoln, R.F. (1981a) Sexual maturation in male triploid plaice *(Pleuronectes platessa)* and plaice × flounder *(Platichthys flesus)* hybrids. *Journal of Fish Biology*, **19**, 415–26.

Lincoln, R.F. (1981b) Sexual maturation in female triploid place *(Pleuronectes platessa)* and plaice × flounder *(Platichthys flesus)* hybrids. *Journal of Fish Biology*, **19**, 499–507.

Lincoln, R.F. and Scott, A.P. (1983) Production of all-female triploids in rainbow trout. *Aquaculture*, **30**, 375–80.

Lincoln, R.F., Aulstad, D. and Grammeltvedt, A. (1974) Attempted triploid induction in Atlantic salmon *(Salmo salar)* using cold shocks. *Aquaculture*, **4**, 287–97.

Linhart, O., Kvasnicka, P., Kanka, J. and Pipota, J. (1986) The first results of induced gynogenesis by retention of the second polar body in common tench *(Tinca tinca* L.) *Bulletin VURH Vodnany*, **1986**, **4**, 3–8.

Liu, H., Yi, Y. and Chen, H. (1987) The birth of the androgenetic homozygous diploid loach *(Misgurnus anguillicaudatus)*. *Acta Hydrobiologica Sinica*, **11**, 246–54.

Liu, S. (1987) Studies on cytogenetics of *Ctenopharyngodon idella, Megalobrama terminalis* and their triploid F_1 hybrid. *Acta hydrobiologica sinica*, **11**, 58–65.

Lou, Y.D. and Purdom, C.E. (1984a) Diploid gynogenesis induced by hydrostatic pressure in rainbow trout eggs fertilized by U-V irradiated spermatozoa. *Journal of Fish Biology*, **24**, 665–70.

Lou, Y.D. and Purdom, C.E. (1984b) Polyploidy produced by hydrostatic pressure in rainbow trout. *Journal of Fish Biology*, **25**, 345–51.

Lowe, T.P. and Larkin, J.R. (1975) Sex reversal in *Betta splendens* Regan with emphasis on the problem of sex determination. *Journal of Experimental Zoology*, **191**, 25–31.

Lucas, G.A. (1968) Factors affecting sex determination in *Betta splendens*. *Genetics*, **60**, 199–200.

Macaranas, I.M., Taniguchi, N., Pante, M.J.R., Capili, J.B. and Pullin, R.S.V. (1986) Electrophoretic evidence for extensive hybrid gene introgression into commercial *Oreochromis niloticus* (L.) stocks in the Philippines. *Aquaculture and Fisheries Management*, **17**, 249–58.

Maceina, J.J. and Murphy, B.R. (1989) Differences in otolith morphology among two subspecies of largemouth bass and their F_1 hybrid. *Transactions of the American Fisheries Society*, **118**, 573–5.

McGeachin, R.B., Robinson, E.H. and Neill, W.H. (1987) Effect of feeding high levels of androgens on the sex ratio of *Oreochromis aureus*. *Aquaculture*, **61**, 317–21.

McIntyre, P. (1961) Crossing over within the macrommelanophore gene in the platyfish, *Xiphophorus maculatus*. *American Naturalist*, **95**, 323–4.

McKay, L.R. and Gjerde, B. (1986) Genetic variation for spinal deformity in Atlantic salmon, *Salmo salar*. *Aquaculture*, **52**, 263–72.

McKenzie, M.D. (1970) First record of albinism in the hammerhead shark *Sphyrna lewini* (Pisces: Sphyrnidae). *Journal of the Elisha Mitchell Scientific Society*, **86**, 35–7.

Maclean, N. and Penman, D. (1990) The application of gene manipulation to aquaculture. *Aquaculture,* **85,** 1–20.

Maclean, N., Penman, D. and Talmar, S. (1987a) Introduction of novel genes into the rainbow trout, in *Selection, Hybridization and Genetic Engineering in Aquaculture,* Vol.2 (ed. K. Tiews), Heenemann, Berlin, pp. 325–33.

Maclean, N., Penman, D. and Zhu, Z. (1987b) Introduction of novel genes into fish. *Bio/Technology,* **5,** 257–61.

McPhail, J.D. and Jones, R.L. (1966) A simple technique for obtaining chromosomes from teleost fishes. *Journal of the Fisheries Research Board of Canada,* **23,** 767–8.

Mair, G.C., Penman, D.J., Scott, A., Skybinsky, D.O.F. and Beardmore, J.A. (1987) Hormonal sex reversal and the mechanisms of sex determination in Oreochromis, in *Selection, Hybridization and Genetic Engineering in Aquaculture* (ed. K. Tiews), Heenemann, Berlin, pp. 301–12.

Maitland, J.R.G. (1887) *The History of Howietoun: the Fish Cultural Work,* Edinburgh University Press, Edinburgh.

Majumdar, K.C. and McAndrew, B.J. (1983) Sex ratios from interspecific crosses within the tilapia, in *Proceedings of the International Conference on Tilapias in Aquaculture* (eds L. Fishelson and Z. Yaron), Tel Aviv University, pp 261–9.

Majumdar, K.C. and McAndrew, B.J. (1986) Relative DNA content of somatic nuclei and chromosomal studies in three genera, *Tilapia, Sarotherodon* and *Oreochromis* of the tribe Tilapiini (Pisces, Cichlidae). *Genetica,* **68,** 175–88.

Maniatis, T., Fritsch, E.F. and Sambrook, J. (1982) *Molecular Cloning: a Laboratory Manual,* Cold Spring Harbor Laboratory, Cold Spring Harbor, NY.

Manning, M.J., Tatner, M.F. and Secombes, C.J. (convenors) (1987) *Immunology and Disease Control Mechanisms of Fish. Journal of Fish Biology,* **31** (Supp. A), 1–261.

Marr, J.C. and Sprague, L.M. (1963) The use of blood group characteristics in studying subpopulations of fishes. *Special Publication of the International Commission for the North West Atlantic Fisheries,* **4,** 308–13.

Mather, K. (1941) Variation and selection of polygenic characters. *Journal of Genetics,* **41,** 159–94.

Matsui, Y. (1934) Genetical studies of gold-fish of Japan. 1. On the varieties of gold-fish and the variations in their external characteristics. 2. On the Mendelian inheritance of the telescope eyes of gold-fish. 3. On the inheritance of the scale transparency of gold-fish. 4. On the inheritance of caudal and anal fins of gold-fish. *Journal of the Imperial Fisheries Institute* **Tokyo, 30,** 1–96.

Matsumoto, J., Kajishima, T. and Hama, T. (1960) Relation between the pigmentation and pterin derivatives of chromatophores during development in the normal black and transparent scaled types of goldfish (*Carassius auratus*). *Genetics,* **45,** 1178–92.

Matsunaga, H. and Oshiro, T. (1987) Allotriploids between goldfish (♀) and carp (♂) induced by cold stock treatment (preliminary note). Bulletin of the Nansei Regional *Fisheries Research Laboratory,* **21,** 11–5.

Mawdesley-Thomas, L.E. (1971) Neoplasia in fish: a review. *Advances in Pathobiology,* **1,** 88–170.

Mayr, E. (1963) *Animal Species and Evolution.* Oxford University Press, Oxford.

Miller, R.B. (1957) Have the genetic patterns of fishes been altered by introduction or by selective fishing? *Journal of the Fisheries Research Board of Canada,* **14,** 797–806.

Miller, R.R. and Schultz, R.J. (1959) All-female strains of the fishes of the genus *Poeciliopsis. Science, NY,* **130,** 1656–7.

Mir, S., Al-Absy, A. and Krupp, F. (1988) A new natural intergeneric cyprinid hybrid from the Jordan River drainage, with a key to the large barbine cyprinids of the southern Levant. *Journal of Fish Biology,* **32,** 931–6.

Mittwoch, V. (1973) *Genetics of Sex Differentiation.* Academic Press, New York.

Moav, R. and Wohlfarth, G. (1976) Two-way selection for growth rate in the common carp *(Cyprinus carpio* L.) *Genetics,* **82,** 83–101.

Moav, R., Hulata, G. and Wohlfarth, G. (1975) Genetic differences between the Chinese

Moav, R., Hulata, G. and Wohlfarth, G. (1975) Genetic differences between the Chinese and European races of the common carp. I. Analysis of genotype-environment interactions for growth rate. *Heredity, Lond.,* **34**, 323–30.

Moav, R., Wohlfarth, G. and Lahman, M. (1964) Genetic improvement of carp. VI. Growth rate of carp imported from Holland relative to Israeli carp and some crossbred progeny. *Bamidgeh,* **16**, 142–9.

Moller, D. (1968) Genetic diversity in spawning cod along the Norwegian coast. *Hereditas,* **60**, 1–32.

Morgan, T.H. (1911) Random segregation versus coupling in Mendelian inheritance. *Science,* **34**, 49.

Morizot, D.C. and Siciliano, M.J. (1984) Gene mapping in fishes and other vertebrates. in *Evolutionary Genetics of Fishes* (ed. B.J. Turner), Plenum Press, New York, pp. 173–234.

Morizot, D.C., Slaugenhaupt, S.A., Kallman, K.D. and Chakravati, A. (1991) Genetic linkage map of fishes of the genus *Xiphophorus* (Teleostii Poeciliidae). *Genetics,* **127**, 399–410.

Mounib, M.S. (1978) Cryogenic preservation of fish and mammalian sperm. *Journal of Reproduction and Fertility,* **53**, 13–18.

Mounib, M.S., Hwang, P.C. and Idler, D.R. (1968) Cryogenic preservation of Atlantic cod (*Gadus morhua*) sperm. *Journal of the Fisheries Research Board of Canada,* **25**, 2623–32.

Muller, U. and Wolf, U. (1979) Cross-reactivity to mammalian anti- H-Y antiserum in teleostean fish. *Differentiation,* **14**, 185–7.

Münzing, J. (1959) Biologie, Variabilitat und Genetik von *Gasterosteus aculeatus* L. (Pisces) Untersuchungen im Elbegebiet. *Internationale Revue der Gesamten Hydrobiologie* **44**, 317–82.

Münzing, J. (1962) Die Populationen der marinen Wanderform von *Gasterosteus aculeatus* L. (Pisces) an den hollandischen und deutschen Nordseekusten. *Netherlands Journal of Sea Research,* **1**, 508–25.

Myers, J.M., Hershberger, W.K. and Iwamoto, R.N. (1986) The induction of tetraploidy in salmonids. *Journal of the World Aquaculture Society,* **17**, 1–7.

Nagy, A., Beresenyi, M. and Csanyi, V. (1981) Sex reversal in carp (*Cyprinus carpio*) by oral administration of methyltestosterone. *Canadian Journal of Fisheries and Aquatic Sciences,* **38**, 725–8.

Nagy, A., Rajki, K., Horvath, L. and Csanyi, V. (1978) Investigation on carp *Cyprinus carpio* L. gynogenesis. *Journal of Fish Biology,* **13**, 215–24.

Nakamura, M. (1975) Dosage-dependent changes in the effect of oral administration of methyltestosterone on gonadal sex differentiation in *Tilapia mossambica. Bulletin of the Faculty of Fisheries, Hokkaido University,* **26**, 99–108.

Nakamura, M. (1981) Effects of 11-ketotestosterone on gonadal sex differentiation in *Tilapia mossambica. Bulletin of the Faculty of Fisheries, Hokkaido University* **47**, 1323–7.

Nakamura, N. and Kasahara, S. (1955) A study on the phenomenon of the tobi-koi or shoot carp. I. On the earliest stage at which the shoot carp appears. *Bulletin of the Japanese Society of Scientific Fisheries,* **21**, 73–6.

Nakamura, N. and Kasahara, S. (1956) A study on the phenomenon of the tobi-koi or shoot carp. II. On the effect of particle size and quantity of food. *Bulletin of the Japanese Society of Scientific Fisheries,* **21**, 1022–4.

Nakamura, N. and Kasahara, S. (1957) A study on the phenomenon of the tobi-koi or shoot carp. III. On the results of culturing the modal group and the growth of fry reared individually. *Bulletin of the Japanese Society of Scientific Fisheries,* **22**, 674–8.

Nakamura, N. and Kasahara, S. (1961) A study on the phenomenon of the tobi-koi or shoot carp. IV. Effects of adding a small number of larger individuals to the experimental hatches of carp fry and culture density upon the occurrence of shoot carp. *Bulletin of the Japanese Society of Scientific Fisheries,* **27**, 958–62.

Nakamura, M. and Takahashi, H. (1973) Gonadal sex differentiation in *Tilapia mossambica* with special regard to the time of estrogen treatment effective in inducing complete feminization of genetic males. *Bulletin of the Faculty of Fisheries, Hokkaido University* **24**, 1–13.

Nakanishi, T. (1987) Kinetics of transfer of immunity by immune leucocytes and PFC response to HRBC in isogenic ginbuna crucian carp. *Journal of Fish Biology*, **30**, 723–9.

Nakanishi, T. and Onozato, H. (1987) Variability in the growth of isogeneic crucian carp *Carassius gibelio langsdorfii*. *Nippon Suisan Gakkaishi*, **53**, 2099–2104.

Naruse, K., Ijiri, K., Shima, A. and Egami, N. (1985) The production of cloned fish in the medaka (*Oryzias latipes*). *Journal of Experimental Zoology*, **236**, 335–41.

Nei, M. (1975) *Molecular Population Genetics and Evolution*. North-Holland, Amsterdam.

Nicolyukin, N.I. (1966) Some problems of cytogenetics, hybridization and systematics of the Acipenseridae. *Genetika (USSR)*, **5**, 25–7.

Nicolyukin, N.I. (1971) Hybridization of Acipenseridae and its practical significance. *F.A.O/United Nations Development Programme (Technical Assistance) Reports on Fisheries.* **(2926)**, 328–34.

Nogusa, S. (1960) A comparative study of the chromosomes in fishes with particular considerations on taxonomy and evolution. *Memoirs of the Hyogo University of Agriculture, (Biological Series)* **3**, 1–62.

Norton, J. (1967a) Bleeding heart topsail platy. *Tropical Fish Hobbyist*, **15**, 4.

Norton, J. (1967b) True hi-fin lyretail sword. *Tropical Fish Hobbyist*, **16**, 4–9.

Nybelin, O. (1947) Ett fall av X-bunden nedärvning hos *Lebistes reticulatus* (PETERS). *Zoologisch Bidrag från Uppsala*, **25**, 448–54.

OECD (1989) *Aquaculture: Developing a New Industry*. OECD, Paris.

O'Farrell, M.M. and Peirce, R.E. (1989) The occurrence of a gynandromorphic migratory trout, *Salmo trutta* L. *Journal of Fish Biology*, **34**, 327.

Ohno, S., Stenius, C., Faisst, E. and Zenzer, M.T. (1965) Post zygotic chromosomal rearrangements in rainbow trout (*Salmo irideus* Gibbons). *Cytogenetics*, **4**, 117–29.

Ohno, S. (1967) *Sex Chromosomes and Sex-linked Genes*. Springer-Verlag, Berlin.

Ohno, S. (1970) The enormous diversity in genome sizes of fish as a reflection of nature's extensive experiments with gene duplication. *Transactions of the American Fisheries Society*, **99**, 120–30.

Ohno, S. (1974) *Animal Cytogenetics*, **4**: *Chordata. 1. Protochordata, Cyclostomata and Pisces*, Gebruder Borntraeger, Berlin.

Okada, H., Matsumoto, H. and Yamazaki, F. (1979) Functional masculinization of genetic females in rainbow trout. *Bulletin of the Japanese Society for Scientific Fisheries*, **45**, 413–9.

Ollerman, L.K. and Skelton, P.H. (1990) Hexaploidy in yellowfish species (*Barbus*, Pisces, Cyprinidae) from southern Africa. *Journal of Fish Biology*, **37**, 105–15.

Ono, R.D., Williams, J.D. and Wagner, A. (1983) *Vanishing Fishes of North America*. Stonewall Press, Washington, DC.

Onozato, H. (1984) Diploidization of gynogenetically activated salmonid eggs using hydrostatic pressure. *Aquaculture*, **43**, 91–7.

Orr, H.A. (1990) "Why polyploidy is rarer in animals than in plants" revisited. *The American Naturalist*, **136**, 759–70.

Oshiro, T. (1987) Cytological studies on diploid gynogenesis induced in the loach *Misgurnus anguillicaudatus*. *Bulletin of the Japanese Society of Scientific Fisheries*, **53**, 933–9.

Owusa-Frimpong, M. and Nijjhav, B. (1981) Induced sex reversal in *Tilapia nilotica* (Cichlidae) with methyltestosterone. *Hydrobiologia*, **78**, 157–60.

Ozato, K., Kondoh, H., Inohora, H., Iwamatsu, T., Wakamatsu, Y. and Okada, T.S. (1986) Production of transgenic fish: introduction and expression of chicken δ-crystallin gene in medaka embryos. *Cell Differentiation*, **19**, 237–44.

Palmiter, R.D., Brinster, R.L., Hammer, R.E., Trumbauer, M.E., Rosenfeld, M.G., Birnberg, N.C. and Evans, R.M. (1982) Dramatic growth of mice that develop from eggs microinjected with metallothionein-growth hormone fusion genes. *Nature, Lond.*, **300**, 611–15.

Pandian, T.J. and Varadaraj, K. (1987) Techniques to regulate sex ratio and breeding in tilapia. *Current Science*, **56**, 337–43.

Pandian, T.J. and Varadaraj, K. (1990) Development of monosex female *Oreochromis mossambicus* broodstock by integrating gynogenetic technique with endocrine sex reversal. *Journal of Experimental Zoology*, **255**, 88–96.

Pape, A. (1935) Beitrage zur Naturgeschichte von *Pleuronectes pseudoflesus*, eines Bastards, zwischen Scholle und Flunder. *Wissenschaftliche Meeresuntersuchungen der Kommission zur Wissenschaftlichen Untersuchungen der Deutschen Meere, Kiel*, **22**, 53–88.

Pardue, M.L. and Gall, J.G. (1970) Chromosomal localization of mouse satellite DNA. *Science*, **168**, 1356–8.

Park, E.H. and Kang, Y.S. (1979) Karyological confirmation of conspicuous ZW sex chromosomes in two species of Pacific anguilloid fishes (Anguilliformes: Teleostomi) *Cytogenetics and Cell Genetics*, **23**, 33–8.

Parker, H.K., Philipp, D.P. and Whitt, G.S. (1985) Relative developmental success of interspecific *Lepomis* hybrids as an estimate of gene regulatory divergence between species. *Journal of Experimental Zoology*, **233**, 451–66.

Pastene, L.A., Numachi, K. and Tsukamoto, K. (1991) Examination of reproductive success in an amphidromous fish, *Plecoglossus altevalis* (Temmink et Schlegel) using mitochondrial DNA and isozyme markers. *Journal of Fish Biology*, **39**, (Supplement A), 93–100.

Payne, R.H. and Child, A.R. (1971) Geographical variation in the Atlantic salmon. *Nature, Lond.*, **231**, 250–52.

Pechan, P., Shapiro, D.Y. and Tracey, M. (1986) Increased H-Y antigen levels associated with behaviourally-induced female-to-male sex reversal in a coral reef fish. *Differentiation*, **31**, 106–10.

Pechan, P., Wachtel, S.S. and Reinboth, R. (1979) H-Y antigen in the teleost. *Differentiation*, **14**, 189–92.

Penman, D., Beeching, A.J., Penn, S. and Maclean, N. (1990) Factors affecting survival and integration following microinjection of novel DNA into rainbow trout eggs. *Aquaculture*, **85**, 35–50.

Peters, G. (1964) Vergleichende Untersuchungen an drei Subspecies von *Xiphophorus hellier*. *Zeitschrift fur die Zoologische Systematik und Evolutionsforschung*, **2**, 185–271.

Philipp, D.P. and Whitt, G.S. (1991) Survival and growth of northern, Florida and reciprocal F₁ hybrid largemouth bass in Central Illinois. *Transactions of the American Fisheries Society*, **120**, 58–64.

Philipp, D.P., Childers, W.F. and Whitt, G.S. (1983) A biochemical genetic evaluation of the northern and Florida subspecies of the largemouth bass. *Transactions of the American Fisheries Society*, **112**, 1–20.

Philipp, D.P., Childers, W.F. and Whitt, G.S. (1985) Correlations of allele frequencies with physical and environmental variables for populations of largemouth bass, *Micropterus salmoides* (Lacepede). *Journal of Fish Biology*, **27**, 347–65.

Phillips, R.B. and Zajicek, K.D. (1982) Q-band chromosomal polymorphism in lake trout (*Salvelinus namaycush*). *Genetics*, **101**, 227–34.

Phillips, R.B., Zajicek, K.D., Ihssen, P.E. and Johnson, O. (1986) Application of silver staining to the identification of triploid fish cells. *Aquaculture*, **54**, 313–19.

Pollord, D.A. (ed.) (1990) *Proceedings of the Workshop on Introduced and Translocated Fishes and their Ecological Effects*. Bureau of Rural Resources, Canberra, *Proceedings*, **8**, pp. 1–181.

Porter, D.A. and Fivizzani, A.J. (1983) Spontaneous occurrence of a synchronous hermaphrodite in the banded killifish, *Fundulus diaphanus* (Leseur). *Journal of Fish Biology*, **22**, 671–5.

Post, A. (1973) Chromosomes of two fish-species of the genus *Diretmus* (Osteichthyes, Beryciformes: Diretmidae), in *Genetics and Mutagenesis in Fish* (ed. J.H. Schroder), Springer-Verlag, Berlin, pp. 104–11.

Powers, D.A., Lauerman, T., Crawford, D. and DiMichele, L. (1991) The evolutionary significance of genetic variation at enzyme synthesizing loci in the teleost *Fundulus heteroclitus*. *Journal of Fish Biology*, **39** (Supplement A), 169–84.

Probst, E. (1949) Der Blauling-Karpfen. *Allgemeine Fischwirtschaftszeitung*, **74**, 232–8.

Pruginin, Y. and Kanyike, E.S. (1965) Mono-sex culture of *Tilapia* through hybridization, in *The Organisation of African Unity, Scientific and Technical Research Commission, Symposium on Fish Farming, Nairobi*, Organisation of African Unity, Nairobi, pp. 1–3.

Pullin, R.S.V. (1972) The storage of plaice (*Pleuronectes platessa*) sperm at low temperatures. *Aquaculture*, **1**, 279–83.

Purdom, C.E. (1969) Radiation-induced gynogenesis and androgenesis in fish. *Heredity, Lond.*, **24**, 431–44.

Purdom, C.E. (1972a) Genetics and fish farming. *MAFF, Lowestoft, Lab. Leaflet*, **25**, 1–16.

Purdom, C.E. (1972b) Induced polyploidy in plaice (*Pleuronectes platessa*) and its hybrid with the flounder (*Platichthys flesus*). *Heredity, Lond.*, **29**, 11–24.

Purdom, C.E. (1973) Variation in fish, in *Sea Fisheries Research* (ed. F.R. Harden Jones) *Elek Press*, pp. 347–55.

Purdom, C.E. (1975) Breeding the domestic fish, in *Proceedings of the Conference on Fish Farming in Europe*, Oyez International Business Communications, London, pp. 61–8.

Purdom, C.E. (1976) Jumbo rainbows: selective breeding or heavy feeding, in *Two Lakes Eighth Fishery Management Training Course Report*, Janssen Services, London, pp. 110–16.

Purdom, C.E. (1977) Fish cultivation research. *MAFF, Lowestoft, Lab. Leaflet*, **35**, 1–29.

Purdom, C.E. (1978) Fish farming research. *Journal of the Royal Agricultural Society of England*, **139**, 124–9.

Purdom, C.E. (1980) Growth in fish, in *Growth in Animals* (ed. T.L.J. Lawrence), Butterworths, London, pp. 273–85.

Purdom, C.E. (1983) Genetic engineering by the manipulation of chromosomes. *Aquaculture*, **33**, 287–300.

Purdom, C.E. (1984) Atypical modes of reproduction in fish, in *Oxford Reviews of Reproductive Biology*, Vol.6 (ed. J.R. Clarke), Clarendon Press, Oxford, pp. 301–40.

Purdom, C.E. and Lincoln, R.F. (1974) Gynogenesis in hybrids within the Pleuronectidae, in *The Early Life History of Fish* (ed. J.H.S. Blaxter), Springer-Verlag, Berlin, pp. 537–44.

Purdom, C.E., Thompson, D. and Dando, P.R. (1976) Genetic analysis of enzyme polymorphisms in plaice *(Pleuronectes platessa)*. *Heredity, Lond.*, **37**, 193–206.

Purdom, C.E., Thompson, D. and Lou, Y.D. (1985) Genetic engineering in rainbow trout, *Salmo gairdneri* Richardson, by the suppression of meiotic and mitotic metaphase. *Journal of Fish Biology*, **27**, 73–9.

Quillet, E., Chevassus, B., Blanc, J.-M., Kreig, F. and Chourrout, D. (1987) Performances of auto and allotriploids in salmonids. I. Survival and growth in fresh water farming. *Aquatic Living Resources*, **1**, 29–43.

Raesly, R.L., Stauffer, J.R. and Denoncourt, R.F. (1990) Hybridization between *Etheostoma zonale* and *Etheostoma olmstedi* (Teleostei: Percidae) following an introduction event. *Copeia*, **1990**, 584–8.

Rakocinski, C.F. (1980) Hybridization and introgression between *Campostoma oligolepis* and *C. anomalum pullum* (Cypriniformes: Cyprinidae). *Copeia*, **1980**, 584–94.

Ralston, E.M. (1934) A study of the chromosomes of *Xiphophorus*, *Platypoecilus* and *Xiphophorus × Platypoecilus* hybrids during spermatogenesis. *Journal of Morphology*, **56**, 423–32.

Rasch, E.M. and Balsano, J.S. (1974) Biochemical and cytogenetic studies of *Poecilia* from eastern Mexico: II. Frequency, perpetuation, and probable origin in triploid genomes in females associated with *Poecilia formosa*. *Revista de Biologia Tropicale*, **21**, 351–84.

Reinboth, R. (1970) Intersexuality in fishes, in *Hormones and the Environment* (eds G.K. Benson and J.G. Phillips), Cambridge Univ. Press, Cambridge, pp. 515–41.

Reinboth, R. (1988) Physiological problems of teleost ambisexuality. *Environmental Biology of Fishes*, **22**, 249–59.

Ridgeway, G.J. (1966) A complex blood group system in salmon and trout, in *Polymorphismes Biochemiques des Animaux*, European Society for Animal Blood Groups Research, Paris, pp. 361–5.

Ridgeway, G.J. and Utter, F.M. (1964) Salmon serology. *Annual Report of the International North Pacific Fisheries Commission*, **1963**, 149–54.

Rivlin, K., Rachlin, J.W. and Dale, G. (1985) A simple method for the preparation of fish chromosomes applicable to field work, teaching and banding. *Journal of Fish Biology*, **26**, 267–72.

Roberts, F.L. (1970) Atlantic salmon (*Salmo salar*) chromosomes and speciation. *Transactions of the American Fisheries Society*, **99**, 105–11.

Rokkones, E., Alestrom, P., Skjervold, H. and Gautvik, K.M. (1989) Microinjection and expression of a mouse metallothionein human growth hormone fusion gene in fertilized salmonid eggs. *Journal of Comparative Physiology*, **158B**, 751–8.

Romashov, D.D., Belyaeva, V.N., Golovinskaia, K.A. and Prokofieva-Bel'govskaya, A.A. (1961) Radiation disease in fish. *Radiatsionnaya Genetika*, **161**, 247–66.

Rosenthal, H.L. and Rosenthal, R.S. (1950) Lordosis, a mutation in the cyprinodont, *Lebistes reticulatus*. *Journal of Heredity*, **41**, 217–18.

Ross, R.M., Hourigan, T.F., Lutnesky, M.M. and Singh, I. (1990) Multiple sex changes in social groups of a coral-reef fish. *Copeia*, **1990**, 427–33.

Russell, A. (1939) Pigment inheritance in the *Fundulus–Scomber* hybrid. *Biological Bulletin*, **77**, 423–31.

Ruzzante, D.E. and Doyle, R.W. (1990) Behavioural and growth responses to the intensity of intraspecific social interaction among medaka, *Oryzias latipes* (Temminck and Schlegel) (Pisces, Cyprinodontidae). *Journal of Fish Biology*, **37**, 663–73.

Ryman, N.C. (1983) Patterns of distribution of biochemical genetic variation in salmonids: differences between species. *Aquaculture*, **33**, 1–21.

Sadovy, Y. and Shapiro, D.Y. (1987) Criteria for the diagnosis of hermaphroditism in fishes. *Copeia*, **1987**, 136–56.

Saikazumi, M., Moriwaki, K. and Egami, N. (1983) Allozymic variation and regional differentiation in wild populations of the fish *Oryzias latipes*. *Copeia*, **1983**, 311–18.

Sambrook, J.S., Fritsch, E.F. and Maniatis, T. (1989) *Molecular Cloning A Laboratory Manual*, Cold Spring Harbor Laboratory Publications, USA.

Scheel, J.J. (1966) Taxonomic studies of African and Asian toothcarps (Rivulinae) based on chromosome numbers, haemoglobin patterns, some morphological traits and crossing experiments. *Videnskabelige Meddelser fra Dansk Naturhistorisk Forening i Kjobenhaun*, **129**, 123–48.

Scheerer, P.D. and Thorgaard, G.H. (1987) Performance and developmental stability of triploid tiger trout (brown trout × brook trout) *Transactions of the American Fisheries Society*, **116**, 92–7.

Schmidt, J. (1920) Racial investigations. IV. The genetic behaviour of a secondary sexual character. *Comptes Rendus des Travail du Laboratoire de Carlsberg*, **14**, No 8, 1–12.

Schroder, J.H. (1964) Genetische Untersuchungen an domestizierten Stammen der Gattung *Mollienesia* (Poeciliidae) *Zoologische Beitrage*, **10**, 369–73.

Schroder, J.H. (1974) *Vererbungslehre fur Aquarianer*, W. Keller, Stuttgart.

Schultz, R.J. (1963) Stubby, a hereditary vertebral deformity in the viviparous fish *Poeciliopsis prolifica*. *Copeia*, **1963**, 325–30.

Schultz, R.J. (1973a) Origin and synthesis of a unisexual fish, in *Genetics and Mutagenesis of Fish* (ed. J.H. Schroder), Springer-Verlag, Berlin, pp. 207–22.

Schultz, R.J. (1973b) Unisexual fish: laboratory synthesis of a species. *Science, NY,* **179**, 180–81.

Schultz, R.J. and Kallman, K.D. (1968) Triploid hybrids between the all-female teleost *Poecilia formosa* and *Poecilia sphenops*. *Nature, Lond.,* **219**, 280–82.

Schwartz, F.J. (1972) World literature to fish hybrids with an analysis by family, species and hybrid. *Publications of the Gulf Coast Research Laboratory Museum,* **3**, 1–328.

Scott, A.P. and Baynes, S.M. (1980) A review of the biology, handling and storage of salmonid spermatozoa. *Journal of Fish Biology,* **17**, 707–39.

Scott, A.P., Baynes, S.M., Skarphedinson, O. and Bye, V.J. (1984) Control of spawning time in rainbow trout, *Salmo gairdneri*, using constant long daylength. *Aquaculture,* **43**, 225–33.

Sedgwick, S.D. (1990) *Trout Farming Handbook*, 5th edn, Fishing News Books, Oxford.

Shaklee, J.B. and Salini, J.P. (1985) Genetic variation and population subdivision in Australian barramundi, *Lates calcarifer* (Bloch). *Australian Journal of Marine and Freshwater Research,* **36**, 203–18.

Shaklee, J.B., Allendorf, F.W., Morizot, D.C. and Whitt, G.S. (1990) Gene nomenclature for protein-coding loci in fish. *Transactions of the American Fisheries Society,* **119**, 2–15.

Shalev, A. and Huebner, E. (1980) Expression of H-Y antigen in the Guppy *(Lebistes reticulatus)*. *Differentiation,* **16**, 81–5.

Shapiro, D.Y. (1988) Behavioural influences on gene structure and other new ideas concerning sex change in fishes. *Environmental Biology of Fishes,* **23**, 283–97.

She, J.X., Autem, M., Kotulas, G., Pasteur, N. and Bonhomme, F. (1987) Multivariate analysis of genetic exchanges between *Solea aegyptica* and *Solea senegalensis* (Teleosts, Soleidae). *Biological Journal of the Linnean Society,* **32**, 357–71.

Shelton, W.L. (1986) Broodstock development for monosex production of grass carp. *Aquaculture,* **57**, 311–19.

Siegal, R.A. (1989) The growing influence of aquaculture in U.S. seafood markets: salmon and shrimp. *Aquaculture: A review of recent experience,* OECD Paris, 1989, 296–313.

Silvers, W.K. and Wachtel, S.S. (1977) H-Y antigen: behaviour and function. *Science, NY,* **195**, 956–60.

Simco, B.A., Goudie, C.A., Klar, G.T., Parker, N.C. and Davis, K.B. (1989) Influence of sex on growth of channel catfish. *Transactions of the American Fisheries Society,* **118**, 427–34.

Simpson, E. (1986) The H-Y antigen and sex reversal. *Cell,* **44**, 813–14.

Skaala, O., Dahle, G., Jorstad, K.E. and Nævdal, G. (1990) Interactions between natural and farmed fish populations: information from genetic markers. *Journal of Fish Biology,* **36**, 449–60.

Slaztenenko, E.P. (1957) A list of natural fish hybrids of the world. *Hidrobiologie, Istanbul,* **4B**, 6–27.

Smith, G.R. (1973) Analysis of several hybrid cyprinid fishes from western North America. *Copeia,* **1973**, 395–410.

Smith, G.R. and Stearley, R.F. (1989) The classification and scientific names of rainbow and cutthroat trouts. *Fisheries,* **14**, 4–10.

Smith, P.J. and Jamieson, A. (1980) Protein variation in the Atlantic mackerel *Scomber scombrus*. *Animal Blood groups and Biochemical Genetics,* **11**, 207–14.

Smith, P.J., Francis, R.I.C.C. and McVeagh, M. (1991) Loss of genetic diversity due to fishing pressure. *Fisheries Research,* **10**, 309–16.

Sola, L., Getili, G. and Cataudella, S. (1980) Eel chromosomes: cytotaxonomical relationships and sex chromosomes. *Copeia,* **1980**, 911–13.

Southern, E.M. (1975) Detection of specific-sequences among DNA fragments separated by gel electrophoresis. *Journal of Molecular Biology,* **98**, 503–17.

Sprague, L.M., Holloway, J.R. and Nakashima, L.I. (1963) Studies of the erythrocyte antigens of albacore, big eye, skipjack, and yellowfin tunas and their use in subpopulation identification. *FAO Fisheries Reports*, **(6) 3**, 1381–93.

Spurway, H. (1953) Spontaneous parthenogenesis in a fish. *Nature, Lond.*, **171**, 437.

Spurway, H. (1957) Hermaphroditism with self fertilization, and the monthly extrusion of unfertilized eggs, in the viviparous fish *Lebistes reticulatus*. *Nature, Lond.*, **180**, 1248–51.

Ståhl, G. (1987) Genetic population structure of Atlantic salmon, in *Population Genetics and Fishery Management* (eds N. Ryman and F. Utter), University of Washington Press, Seattle, WA, pp. 121–40.

Stanley, J.G. (1976) Production of hybrid, androgenetic and gynogenetic grass carp and carp. *Transactions of the American Fisheries Society*, **105**, 10–6.

Stanley, J.G. and Sneed, K.E. (1974) Artificial gynogenesis and its application in genetics and selective breeding of fishes, in *The Early Life History of Fish* (ed. J.H.S. Blaxter), Springer Verlag, Berlin, pp. 527–36.

Stebbins, G.L. (1958) Hybrid inviability, weakness and sterility. *Advances in Genetics*, **9**, 147–215.

Sterba, G. (1962) *Freshwater Fishes of the World*, (transl. and rev. D.W. Tucker), Vista Books, London, 878.

Stet, R.J.M., Kaastrup, P., Egberts, E. and Van Muiswinkel, W.B. (1990) Characterization of new immunogenetic markers using carp alloantisera: evidence for the presence of Major Histocompatibility Complex (MHC) molecules. *Aquaculture*, **85**, 119–24.

Stevenson, M.M. and Buchanan, T.M. (1973) An analysis of hybridization between the Cyprinodont fishes *(Cyprinodon variegatus* and *C. elegans)*. *Copeia*, **1973**, 682–92.

Stoss, J. (1983) Fish gamete preservation and spermatozoan physiology, in *Fish Physiology*, Vol. 9B (eds W.S. Hoar, D.J. Randall and E.M. Donaldson), Academic Press, London, pp. 305–50.

Stoss, J., Buyukhatipoglu, S. and Holtz, W. (1978) Short term and cryopreservation of rainbow trout (*Salmo gairdneri* Richardson) sperm. *Annales de Biologie Animale Biochimie et Biophysique*, **18**, 1077–82.

Streisinger, G., Walker, C., Dower, N., Knauber, D. and Singer, F. (1981) Production of clones of homozygous diploid zebra fish *(Brachydanio rerio)*. *Nature, Lond.*, **291**, 293–6.

Sumner, A.T., Evans, H.J. and Buckland, R. (1971) New technique for distinguishing between human chromosomes. *Nature, Lond., New Biol.*, **232**, 31–2.

Sutton, W.S. (1902) Chromosomes in heredity. *Biological Bulletin*, **4**, 231–51.

Suzuki, A., Morio, T. and Mimoto, K. (1959) Serological studies of the races of tuna. II. Blood group frequencies of the albacore in Tg system. *Report of the Nankai Regional Fisheries Research Laboratory*, **11**, 17–23.

Svardson, G. (1945) Chromosome studies on Salmonidae. *Meddelanden fran Statens Undersokningsforsogsanstalt for Sotvattenfisket*, **23**, 1–15.

Swarup, H. (1959) Production of triploidy in *Gasterosteus aculeatus* (L). *Journal of Genetics*, **56**, 129–42.

Swofford, D.L. and Selander, R.B. (1981) BIOSYS-1: a FORTRAN programme for the comprehensive analysis of electrophoretic data in population genetics and systematics. *Journal of Heredity*, **72**, 282–3.

Tait, J.S. (1970) A method of selecting trout hybrids (*Salvelinus fontinalis S. namaycush*) for ability to retain swimbladder gas. *Journal of the Fisheries Research Board of Canada*, **27**, 39–45.

Takeuchi, K. (1976) 'Wavy-fused' mutants in the medaka, *Oryzias latipes*. *Nature, Lond.*, **121**, 866–7.

Talent, L.G. (1973) Albinism in embryo gray smoothound sharks, *Mustelus californicus*, from Elkhorn Slough, Monterey Bay, California. *Copeia*, **1973**, 595–7.

Tamadachi, M. (1990) *The Cult of the Koi*. TFH Publications, Jersey City, NJ.

Tang, F., Chang, S.T.H. and Lofts, B. (1974) Effect of steroid hormones on the process of natural sex reversal in the rice field eel, *Monopterus albus* (Zuiew). *General and Comparative Endocrinology*, **24**, 227–41.

Taniguchi, N., Kijima, A. and Fukai, J. (1987) High heterozygosity at *Gpi-1* in gynogenetic diploids and triploids of ayu *Plecoglossus altivelis*. *Nippon Suisan Gakkaishi*, **53**, 717–20.

Taniguchi, N., Kijima, A., Fukai, J. and Inada, Y. (1986) Conditions to induce triploid and gynogenetic diploid in ayu (*Plecoglossus altivelis*). *Bulletin of the Japanese Society of Scientific Fisheries*, **52**, 49–53.

Taniguchi, N., Seki, S., Fukai, J. and Kijima, A. (1988) Induction of two types of gynogenetic diploids by hydrostatic pressure shock and verification by genetic marker in ayu. *Nippon Suissan Gakkaishi*, **54**, 1483–91.

Tave, D. (1988) Body color in rainbow trout. *Aquaculture Magazine.*, **14 (3)**, 65–6.

Tayamen, M.M. and Shelton, W.L. (1978) Inducement of sex reversal in *Sarotherodon niloticus* (Linnaeus). *Aquaculture*, **14**, 349–54.

Teichert-Coddington, D.R. and Oneal Smitherman, R. (1988) Lack of response by *Tilapia nilotica* to mass selection for rapid early growth. *Transactions of the American Fisheries Society*, **117**, 297–300.

Thompson, D. (1985) Genetic identification of trout strains. *Aquaculture*, **46**, 341–51.

Thompson, D. and Iliadou, K. (1990) A search for introgressive hybridization in the rudd, *Scardinius erythrophthalmus* (L.) and the roach, *Rutilus rutilus* (L.). *Journal of Fish Biology*, **37**, 367–73.

Thompson, D. and Scott, A.P. (1984) An analysis of recombination data in gynogenetic diploid rainbow trout. *Heredity, Lond.*, **53**, 441–52.

Thorgaard, G. (1977) Heteromorphic sex chromosomes in male rainbow trout. *Science, NY*, **196**, 900–902.

Thorgaard, G. (1983a) Chromosomal differences among rainbow trout populations. *Copeia*, **1983**, 650–62.

Thorgaard, G.H. (1983b) Chromosome set manipulation and sex control in fish, in *Fish Physiology* Vol.9B (eds W.S. Hoar, D.J. Randall and E.M. Donaldson), Academic Press, New York, pp. 405–34.

Thorgaard, G.H., Rabinovitch, P.S., Shen, M.W., Gall, G.A.E., Propp, J. and Utter, F.M. (1982) Triploid rainbow trout identified by flow cytometry. *Aquaculture*, **29**, 305–9.

Thorgaard, G.H., Sheerer, P.D., Hershberger, W.K. and Myers, J.M. (1990) Androgenetic rainbow trout produced using sperm from tetraploid males show improved survival. *Aquaculture*, **85**, 215–21.

Thorpe, J.E. (1985) Early maturity in male Atlantic salmon. *Scottish Fisheries Bulletin*, **42**, 15–17.

Todd, T.N. (1986) Occurrence of white bass-white perch hybrids in Lake Erie. *Copeia*, **1986**, 196–9.

Tripathy, N.K. and Das, C.C. (1980) Chromosomes in three species of Asian catfish. *Copeia*, **1980**, 916–18.

Tropical Fish Hobbyist, **1–39** (1952-1991).

Tucker, C.C. and Robinson, E.H. (1990) *Channel Catfish Farming Handbook*, Van Nostrand Reinhold, New York.

Turner, B.J. (1991) Repetitive DNA sequences and the divergence of fish populations: some hopeful beginnings. *Journal of Fish Biology*, **39** (Supplement A), 131–42.

Ubisch, L. Von (1955) Uber die Zahl der Flossenstrahlen bei *Pleuronectes platessa* und der Bastard und Ruckkreutzungen zwischen beiden Arten. *Zoologische Anzeiger*, **151**, 75–88.

Ueda, T. and Ojima, Y. (1984a) Sex chromosomes in the rainbow trout (*Salmo gairdneri*). *Bulletin of the Japanese Society of Scientific Fisheries*, **50**, 1499–1504.

Ueda, T. and Ojima, Y. (1984b) Sex chromosomes in the Kokanee salmon *Oncorhynchus nerka*. *Bulletin of the Japanese Society of Scientific Fisheries*, **50**, 1495–8.

Utter, F.M., Compton, D., Grant, S., Milner, G., Seeb, J. and Wishard, L. (1980) Population structures of indigenous salmonid species of the Pacific Northwest, in

Salmonid Ecosystems of the North Pacific (eds W.J. McNeil and D.C. Himsworth), Oregon State University Press, pp. 285–304.

Utter, F.M., Mighell, J.L. and Hodgins, H.O. (1973) Inheritance of biochemical variants in three species of Pacific salmon and rainbow trout. *Annual Report of the International North Pacific Fisheries Commission*, **1971**, 97–100.

Utter, F.M., Ridgeway, G.J. and Hodgkins, H.O. (1964) Use of plant extracts in serological studies of fish. *Special Scientific Reports of the U.S. Fisheries and Wildlife Service*, **472**, 1–17.

Valenti, R.J. (1975) Induced polyploidy in *Tilapia aurea* (Steindachner) by means of temperature shock treatment. *Journal of Fish Biology*, **7**, 519–28.

Vanderplank, F.L. (1972) Colour in goldfish and koi. *The Aquarist and Pondkeeper*, **36**, 386–7.

Varadaraj, K. (1989) Feminization of *Oreochromis mossambicus* by the administration of diethylstilbestrol. *Aquaculture*, **80**, 337–41.

Varadaraj, K. (1990) Dominant red colour morphology used to detect paternal contribution in batches of *Oreochromis mossambicus* (Peters) gynogens. *Aquaculture and Fisheries Management*, **21**, 163–72.

Varadaraj, K. and Pandian, T.J. (1989) First report on production of supermale tilapia by integrating endocrine sex reversal with gynogenetic technique. *Current Science*, **58**, 434–41.

Vasetskii, S.G. (1967) Changes in the ploidy of sturgeon larvae induced by heat treatment of eggs at different stages of development. *Doklady Academii Nauk SSSR, Biological Sciences*, **172**, 23–6.

Vasetskii, S.G., Betina, M.I. and Kondrat'eva, O.T. (1984) Obtaining triploids by exerting hydrostatic pressure on fertilized eggs of *Misgurnus fossilis*. *Doklady Academii Nauk SSSR, Biological Sciences*, **279**, 667–9.

Vasiliev, V.P., Makeeva, A.P. and Ryabov, I.N. (1975) Triploidy of hybrids of carp with other representatives of the family Cyprinidae. *Genetika (USSR)*, **11**, 49–56.

Vasiliev, V.P., Vasilieva, E.D., Ivanov, V.N. and Bulatov, K.V. (1984) An analysis of natural hybridization in the Centracanthidae (Pisces, Perciformes) from the Black Sea. *Zoologichesky Zhurnal*, **63**, 554–62.

Vasil'yeva, Ye.D. (1990) On morphological divergence of the gynogenetic and bisexual forms of the goldfish, *Carassius auratus* (Cyprinidae, Pisces). *Journal of Ichthyology*, **30**, 22–37.

Vaupel, J. (1929) The spermatogenesis of *Lebistes reticulatus*. *Journal of Morphology*, **47**, 555-87.

Verspoor, E. (1988) Widespread hybridization between native Atlantic salmon, *Salmo salar*, and introduced brown trout, *S. trutta*, in eastern Newfoundland. *Journal of Fish Biology*, **32**, 327–34.

Verspoor, E. and Hammar, J. (1991) Natural hybridization and introgression in fishes: the biochemical evidence. *Journal of Fish Biology*, **39** (Supplement A), 307–34.

Vielkind, J., Haas-Andela, H., Vielkind, V. and Anders, F. (1982) The induction of a specific pigment cell type by total genomic DNA injected into the neural crest region of fish embryos of the genus *Xiphophorus*. *Molecular and General Genetics*, **185**, 379–89.

Vitturi, R., Mazzola, A., Macaluso, M. and Catalano, E. (1986) Chromosomal polymorphism associated with Robertsonian fusion in *Seriola dumerili* (Risso, 1810) (Pisces: Carangidae). *Journal of Fish Biology*, **29**, 529–34.

Wachtel, S.S., Ohno, S., Koo, G.C. and Boyce, E.A. (1975) Possible role for H-Y antigen in the primary determination of sex. *Nature, Lond.*, **257**, 235–6.

Wallbrunn, H.M. (1957) Genetics of the Siamese fighting fish, *Betta splendens*. *Genetics*, **43**, 289–98.

Ward, R.D. and Beardmore, J.A. (1977) Protein variation in the plaice, *Pleuronectes platessa* L. *Genetical Research, Cambridge*, **30**, 45–62.

Wheat, T.E. and Whitt, G.S. (1973) Linkage relationships of six enzyme loci in interspecific sunfish hybrids (genus *Lepomis*). *Genetics*, **74**, 343–50.

Whitt, G.S. (1981) Developmental genetics of fishes: isozymic analyses of differential gene expression. *American Zoologist*, **21**, 549–72.

Wiberg, U. (1982) Serological cross-reactivity to rat anti- H-Y antiserum in the female European eel (*Anguilla anguilla*). *Differentiation*, **21**, 206–8.

Winemiller, K.O. and Taylor, D.H. (1982) Inbreeding depression in the convict cichlid, *Cichlasoma nigrofasciatum* (Baird and Girard). *Journal of Fish Biology*, **21**, 399–402.

Winge, O. (1922a) A peculiar mode of inheritance and its cytological explanation. *Journal of Genetics* **12**, 137–44.

Winge, O. (1922b) One-sided masculine and sex-linked inheritance in *Lebistes reticulatus*. *Journal of Genetics* **12**, 145–62.

Winge, O. (1927) The location of eighteen genes in *Lebistes reticulatus*. *Journal of Genetics*, **18**, 1–43.

Winge, O. (1934) The experimental alteration of sex chromosomes into autosomes and vice versa, as illustrated by *Lebistes*. *Compte Rendu des Travaux du Laboratoire de Carlsberg, Serie Physiologique*, **21**, 1–49.

Winge, O. and Ditlevson, E. (1938) A lethal gene in the Y chromosome of *Lebistes*. *Compte Rendu des Travaux du Laboratoire de Carlsberg, Serie Physiologique*, **22**, 205–11.

Winge, O. and Ditlevson, E. (1946) Colour inheritance and sex determination in *Lebistes*. *Compte Rendu des Travaux du Laboratoire de Carlsberg, Serie Physiologique*, **24**, 227–48.

Wirgin, I.I. and Maceda, L. (1991) Development and use of striped bass-specific RFLP probes. *Journal of Fish Biology*, **39** (Supplement A), 159–67.

Wohlfarth, G.W. (1977) Shoot carp. *Bamidgeh*, **29**, 35–6.

Wohlfarth, G.W., Lahman, M. and Moav, R. (1963) Genetic improvement of carp. IV. Leather and line carp in fish ponds of Israel. *Bamidgeh*, **15**, 3–8.

Wohlfarth, G.W. and Hulata, G.I. (1983) Applied genetics of tilapias. ICLARM Studies and Reviews, **6**, International Centre for Living Aquatic Resources Management, Manila, 26 pp.

Wohlfarth, G.W., Moav, R. and Lahman, M. (1961) Genetic improvements of carp. III. Progeny tests for differences in growth rate. *Bamidgeh*, **13**, 40–54.

Woiwode, J.G. (1977) Sex reversal of *Tilapia zillii* by injestion of methyltestosterone. *Technical Paper Ser Bureaux of Fisheries and Aquatic Resources* (Philippines) **1**, 1–5.

Wolters, W.R., Libey, G.S. and Chrisman, C.L. (1981) Induction of triploidy in channel catfish. *Transactions of the American Fisheries Society*, **110**, 310–2.

Wood, A.B. and Jordan, D.R. (1987) Fertility of roach × bream hybrids, *Rutilus rutilus* (L.)× *Abramis brama* (L.), and their identification. *Journal of Fish Biology*, **30**, 249–61.

Wootton, R.J. (1976) *The Biology of Sticklebacks*. Academic Press, London.

Wright, J.E.jun. and Atherton, L.M. (1970) Polymorphism for LDH and transferrin loci in brook trout populations. *Transactions of the American Fisheries Society*, **99**, 179–92.

Yamamoto, T. (1953) Artificially induced sex-reversal in the genotypic males of the medaka (*Oryzias latipes*). *Journal of Experimental Zoology*, **123**, 571–94.

Yamamoto, T. (1955) Progeny of artificially induced sex-reversals of the male genotype (XY) in the medaka (*Oryzias latipes*) with special reference to the YY male. *Genetics, Princeton*, **40**, 406–19.

Yamamoto, T. (1958) Artificial induction of functional sex-reversal in genotypic females of the medaka (*Oryzias latipes*). *Journal of Experimental Zoology*, **137**, 227–64.

Yamamoto, T. (1964) Linkage map of sex chromosomes in the medaka (*Oryzias latipes*). *Genetics*, **50**, 59–64.

Yamamoto, T. (1969a) Inheritance of albinism in the medaka, *Oryzias latipes*, with special reference to gene interaction. *Genetics*, **62**, 797–809.

Yamamoto, T. (1969b) Sex differentiation, in *Fish Physiology*, Vol.3 (eds W.S. Hoar and D.J. Randall), Academic Press, New York, pp. 117–175.

Yamazaki, F. (1983) Sex control and manipulation in fish. *Aquaculture*, **33**, 329–54.

Yeager, D.M. (1985) Creation of a hybrid striped bass fishery in the Escambia River, Florida. *North American Journal of Fisheries Management*, **5**, 389–92.

Yoon, S.J., Hallerman, E.M., Gross, M.L., Liu, Z., Schneider, J.R., Faras, A.J., Hacket, P.B., Kapuscinski, A.R. and Guise, K.S. (1990) Transfer of the gene for neomycin resistance into goldfish, *Carassius auratus. Aquaculture*, **85**, 21–33.

Yu, H. (1982) A cytological observation on gynogenesis of crucian carp (*Carassius auratus gibelio*). *Acta Hydrobiologica Sinica*, **7**, 481–7.

Zhang, J. (1985) A study of reciprocal cross hybrids and backcross hybrids of *Cyprinus carpio var. wuyuanensis* with *C. carpio yuankiang* and the economic benefit in F_2. *Journal of Fisheries of China*, **9**, 375–82.

Zhu, Z., Li, G., He, L. and Chen, S. (1985) Novel gene transfer into the fertilized eggs of goldfish (*Carassius auratus* L. 1758). *Journal of Applied Ichthyology*, **1**, 31–4.

Zohar, Y., Abraham, M. and Gordin, H. (1978) The gonadal cycle of the captivity-reared hermaphroditic teleost *Sparus aurata* (L.) during the first two years of life. *Annales de Biologie Animale, Biochimie et Biophysique*, **18**, 877–82.

Index

Abramis brama 163
Acipenser guldenstadti × *A. baeri* 173
Acipenser ruthenus 172
Acrosome 22
Albinism
   in the Guppy 28
   inheritance 26–8
   in the medaka 27
   in the stickleback 29
Alburnus albidus 162
Alburnus alburnus 162
Allotriploids 220–1
Allozymes, *see* Isozymes
Ameca splendens
   chromosomes 116
   internal broods 16
   sex ratio 137
American stickleback plate morphs 50–3
Androgenesis 204, 212–13
Androgenesis and YY males 196
Anguilla anguilla 152, 157
Anthias squamipinnis
   hermaphroditism 184
   H-Y antigenin 157
   social behaviour 182
Anti freeze protein 230
Antigen/antibody relationships 86–8
Apeltes quadracus 154
Aphyosemion calliurum 167
Aral Sea 11
Artificial fertilization 166
Artificial insemination 24
Autosomal inheritance 32

Balanced polymorphism 94
Barbus canis 162
Base pairs 112
Bester 172–3
Betta splendens
   colour 233
   putative sex chromosomes 152
   sex ratios 136
   sex reversal 182
Biochemical characteristics 44
Blood agglutination techniques 87
Blood group data from subpopulations
   88–9
Blood groups 86
Body shape 44, 54–5
Brachydanio rerio
   diploid gynogenesis 208
   mitotic gynogenesis 210
   sex determination 146
Branchiostoma belcheri 149
Brownbows 172

Campostoma oligolepis 164
Capoeta damascina 162
Carassius auratus
   female-only forms 189
   gene transfer in 227, 230
   ornamental strains 54, 233
Carassius auratus gibelio 189
Carassius auratus langsdorfi 189
Carassius carassius 169
Cetorhinus maximus 1
Chanos chanos 3

Charles Darwin 25
Chiasma *see* Crossing over
Chromatids 20
Chromosome
  arm structures 112
  C-bands 117
  engineering 204
  G-bands 117
  general 17, 25, 110–11
  in humans 117
  number 34, 110
  polymorphisms 124
  primitive 121
  Q-bands 117
  staining techniques 117–19
  variation 122–7
Cichlasoma nigrofasciatus 103
Clines 106
Clones
  in female only species 75, 181, 186
  production by gynogenesis 210
Clupea harengus 16, 88
Co-dominant alleles 59
Coadaptation and genotype/phenotype
  code 176–7
Coadapted gene pools 105
Colour patterns of *Xiphophorus*
  maculatus 36–43
Colour pigments 26
Colour polymorphism in *Poecilia* 32
Common carp 241
Conservation and fish introduction 243
Coris julis 157, 182
Cottis bairdii 149
Crossing over
  general 20, 41, 114–15
  in the Guppy 128–31
  sex chromosomes 141
Cryopreservation 22–3
Ctenopharyngodon idella 146, 166
Cyprinodon elegans 163
Cyprinodon variegatus 163
Cyprinus carpio
  farming 2, 6, 23
  gynogenesis in 146
  hybridization 169, 189
  ornamental forms *see also* Koi
  scale patterns 44–7

Cytological methods 111–13, 115–19

Degree of heterozygosity 95–7
Determinate growth 69
Developmental homeostasis 176
Diapause 24
Dicentrarchus labrax 6
Diploid 20, 110
Diploid gynogenesis
  by chemical treatment 208–9
  by cold shock 208–9
  and chromosome mapping 131
  by heat shock 208–9
  by hydrostatic pressure 208–9
  meiotic 205–10
  mitotic 210–12
  use for inbreeding 209–10
Diretmus argentus 149
Disease resistance 82–3
DNA
  fingerprinting 100–1
  the genetic material 112
  helix 25
  replication of introduced 227
Domesticated strains of trout 243
Duplicate loci 62, 91

Egg production 5, 16–21
Electrophoretic variation 57–63,
  89–94
Engraulis encrasicolus 88
Environmental variance 68
Enzyme Council numbers 91
Epistasis 29
Euchromatin 111–12
Euheterosis 159

F_1 hybrids and inbreeding 101
Female-only
  artificial forms 189
  breeding in XX/XY system 194–5
  species 185–6
  UK production survey 202
Feminization
  in cichlids 201
  in salmonids 200
  with oestrogens 195
Fin shape 55–7, 239

First polar body 20
Fish cultivation *see* Fish farming
Fish farming
    aesthetic 8
    extensive 6–7
    intensive 4–6
    by natural productivity 7
    stock enhancement 107
    world trends 4
Fishing
    methods 8–10
    statistics 8
Food conversion efficiency 81
Fundulus heteroclitus 106

Gadus morhua 9, 23, 88, 106
Gambusia affinis 153
Gasterosteus aculeatus 239–40
Gasterosteus aculeatus scale patterns
    47–53
Gasterosteus wheatlandi 154
Gene
    concept 25
    insertion 226–7
    interaction 30
    libraries 224
    pool 53
    simple effects 26
    transfer 223–7
    transfer assessment 227–31
    transfer expression 231–2
Genetic
    distances 94
    drift 102
    load 103
    relatedness 97–8
    variance 68
Genome size 120
Genotypic binomial 66–8
Geographic strains of salmonids 242–3
Ginbuna carp 75
Golden trout 172
Goldfish
    colour varieties 234
    fancy strains 234–5
    scale less phenotypes 46–7
Gonad 13, 14
Gonochorism 12, 178

GESAMP 10–11
Growth hormone 231–2
Growth hormone gene 225
Growth in rainbow trout 72
Growth rate
    in Atlantic salmon 74
    and environment 72
    in *Gambusia affinis* 75
    genetics 69–76
    in tilapia 75
    selection programmes 70–5
Guppy
    colour patterns 35
    varieties 238–9
    veiltails 56
Guppy Reverend R. J. L. 32
Gynogenesis
    and chromosome engineering 204
    and chromosome mapping 131
    meiotic 206–10
    mitotic 210–12
    and sex ratio control 196

H-Y antigen
    in sex determination 156–7
    in hermaphrodites 184
Habitat change 2, 10–11
Haemoglobin 90
Haldane, J. B. S. 145
Haplochromis burtoni 157
Haploid 20, 110, 152, 206–7
Hardy–Weinberg law 93–4
Heritability 68–9, 82
Hermaphroditism
    abnormal 184–5
    cyclical 179
    experimental production of 185
    normal 12, 178–84
    protandrous 179
    protogynous 179
    in Serranidae and Sparidae 179–80
Hertwig, O. 204
Heteroagglutination 87–8
Heterochromatin 111
Heteroplasmy 100
Heterosis
    in F_1 hybrids 102
    in species crosses 159, 161

Heterosis (continued)
    in racial crosses 174
    in strain crosses 103
Hierarchy and growth rate 75–6
Hippoglossus hippoglossus 105, 208
Holocanthus ciliaris 164
Holocanthus isabelita 164
Homozygosity
    by descent 102
    in gynogenesis 210–12
Hubbs, Carl 159
Huso huso 172
Hybrid
    intermediacy 161
    meristics and morphometrics 158,
        161–2
    sterility 162–4
    vigour 102, 162
Hybrid crosses
    in bass 169–70
    in *Cyprinus carpio* 174
    index 161
    between races 173
    reciprocal 171
    in salmonids 170–2
    of sturgeon 172–3
    in trout strains 174
Hybridization
    artificial 166–77
    application of 169–75
    and chromosome mapping 132
    and development genetics 167–8
    and environment 158, 164–5
    and introductions 165
    introgressive 160, 162, 163–4, 175–7
    in nature 159–65
    and selection 168–9
    and taxonomy 166–7
    and tumour formation 167
Hybridogenesis 188

Ictalurus nebulosus natural agglutinins 86
Ictalurus punctatus 7, 230
Ictalurus punctaus growth rate 73
Idus idus 27
Immune mechanisms 63
Inactivation of spermatozoa 207–8
Inbred lines 103

Inbreeding
    depression 101
    by mating relatives 101–4, 173
    and skeletal deformity 54
Indeterminate growth 69
Intercalary heterochromatin 101
Interference 130, 132
Introductions and introgression 100,
        104, 107
Introgressive hybridization 104, 107,
        163–4
Isoagglutination 87
Isogenic clones 75
Isozyme models 90–4
Isozymes
    patterns 89–91
    tissue expression 62–3
Israeli carp breeding school 70

Karyotypes 119
Killifish letters 167
Koi
    crosses 237–8
    fin shape 238
    selection in 83
    varieties 235–8

Lactate dehydrogenase 62
Lake Friant 53
Lake Wapata 52
Lamarck, J. B. 25
Landsteiner 85
Late replicating DNA 117
Lates calcarifer 108
Leather carp 45
Lepomis cyanellus 159
Leuciscus cephalus 162
Linkage and chromosome maps 127–33
Linkage disequilibrium 94
Linkage groups
    in salmonids 133
    in Xiphrophorus hybrids 132–3
Lordosis 54
Lough Melvin trout 106
Luxuriance 159, 167

Macromelanophore gene transfer 224
Macropodus opercularis 27

Major histocompatibility systems 63, 82
Malacca tilapia hybrids 145
Malate dehydrogenase 61
Male-only breeding in XX-XY system 195–7
Mallotus villosus 100
Masculinization
   of cichlids 198
   of salmonids 198
   and sex determination 143–4
Meiosis 20–2, 114–15
Meiosis avoidance of 188
Melanogrammus aeglifinus 88
Mendel, Gregor 25
Mendelism 26
Meristics 158
Merluccius merluccius 105
Metallothionein gene 225
Metaphase 113
Metaphase spindle 113
Metrical traits 65
Microchromosomes 121, 149, 173
Micropterus salmoides floridanus 173–4
Micropterus salmoides salmoides 173–4
Micropyle 22
Mirror carp 45
Misgurnus anguillicaudatus 205, 212
Mitochondrial DNA 99–100
Mitosis 18–20, 113
Modifier genes 46
Mogurnda obscura 149
Molecular genetics 26
Monopterus alba 181, 183
Monosex culture *see* Sex ratio control
Morgan, T. H. 127
Morone americana 162–3
Morone chrysops 162
Morone saxatilis 95, 105
Morphometrics 158
Mugil cephalus 23
Muller, H.J. 103
Mullerian duct 181
Mullus surmelatus 2
Multiple alleles 40
Mutation rates 85
Mutations 85, 95

Nei's genetic distance 97–8

Nishikigoi see Koi
Nombre fondamental 112
Non-disjunction 42, 114, 122
Nucleolar organizers 117
Null allele 86, 89

Oncorhynchus gorbuscha 169
Oncorhynchus keta 169
Oncorhynchus kisutch 198, 200
Oncorhynchus masou 200
Oncorhynchus mykiss
   feminization 200
   hybrids of 172
   introduction impact of 165
   reproductive success 107
   sex control 198
   sex determination 154
   sperm cryopreservation 23
Oncorhynchus nerka 155
Oncorhynchus tshawytscha 198
Ontogenic succession 91
Oreochromis aureus
   feminization 201
   gynogenesis in 147
   masculinization 198
Oreochromis hornorum 145
Oreochromis macrochir 198
Oreochromis mossambica
   feminization 201
   hybrids 145
   sex determination 147
Oreochromis niloticus 198, 201
Oreochromis zillii 198
Oryzias latipes
   albinism 27
   colour genetics 30–2
   mitotic gynogenesis in 211–12
   selection for aggression 76
   sex linkage in 151
   spinal deformity 54
   transgenics 230
Ovulation 21

Panmixia 93
Parental care 47
Paternal inheritance in the Guppy 138
Penetrance 51
Pericentric inversion 122

Phosphoglucomutase 61
Pigment cells 26–38, 39
Platichthys flesus 164, 215
Platy hifin lyretail 56
Plecoglossus altivelis 100, 207
Pleiotropy 29
Pleuronectes platessa
    electrophoretic polymorphisms 95
    gynogenesis 207
    natural hybrids 164
    population studies 105
    sex determination 146
Poecilia reticulata
    albinism 28
    colour patterns 30, 32–6
    sex linkage 137–8
    Y chromosome 151, 154
Poecilia unisexuals 186–8
Poeciliopsis prolifica 54
Poeciliopsis unisexual complex 188–9
Polygene 50, 66
Polygenes and environment 68
Polygenic inheritance 66–8
Polymorphic 95
Polymorphism
    adaptive 106
    for chromosome banding 127
    for electropheretic loci 95
    for sex chromosomes 139
Polyodon spathula 95
Polyploidy induced 110, 206, 213–22
Polyspermy 22
Population genetics
    applications 104–9
    blood group data 88–9
    concepts 65
    isozyme data 91–4
Population divergence 104–5
Primary oocytes 16
Primordial germ cells 14
Pseudopleuronectes americana 88, 230

Rainbow trout colour 42
Realized heritability 69
Recombination 114, 127
Reproduction 12, 16
Restriction endonucleases 99
Rivulus marmoratus 178, 181, 185

Robertsonian fusion 122
Rutilus rubilio 162
Rutilus rutilus 162–3

Salmo clarkii 164
Salmo salar
    hybrids 170
    mt DNA studies 100
    Nombre fondamental variation 126
Salmo trutta
    anadromous 135
    chromosome number 112
    disease resistant 82–3
    feminization 200
    hybrids 170
    introduction 165
    sympatric populations 106
Salvelinus fontinalis 62, 82, 168
Salvelinus namaycush 168, 171
Satellite DNA 101
Scale and fin characteristics 44–53, 56–67
Scardinius erythrophthalmus 27, 152
Schmidt, Johannes 33, 137
Sea ranching 7–8
Second polar body 21
Secondary sexual characteristics 182, 192
Selection
    disease resistance 82
    fin and body shape 235
    growth rate 69–76
    spawning time 80
Selection in fancy fish 83–4
Selection and fisheries 108
Seiection pressure 68
Self fertilization
    in *Fundulus diaphanus* 184
    in *Rivulus marmoratus* 184
    in *Salmo* and *Oncorhynchus* 185
    in the Guppy, 184
Seriola quinqueradiata 2
Sex chromosome
    cytology 148–56
    maps 141–2
    specialization 131
Sex chromosomes
    crossing over restriction 151
    of deep sea fish 150–2
    and sex determination 134

Sex Chromosomes(continued)
  and sex linkage 31
  of shallow waterfish 152–5
Sex determination
  and androgenesis 146–7
  genetics 134–5, 137–41
  and gynogenesis 146–8
  and hybridization 145–6
  polygenic 136–7
  Sex reversal 143–6
Sex hormones 15
Sex inversion 144
Sex linkage 137–42
Sex ratio 16, 135–6, 192
Sex ratio control
  commercial application 201–3
  methods 194–201
Sex reversal
  hormone induced 197–201
  in hermaphroditism 181–4
Sex-linked inheritance
  historical 31–2
  in the Guppy 34–6
  in the platy 37
Sexual characteristics
  primary 13–14
  secondary 15–16
Sexuality 12
Silver carp unisexuals 189–90
Skeletal abnormalities 54
Solea solea 9
Splake trout 168
Sparus auratus 181, 183
Spawning time 79–80
Spermatozoa production 21
Sphyrna levini 26
Steelhead trout 244
Stem cells 232
Stizostedion species divergence 97
Stock enhancement 7
Stocks and strains
  Atlantic salmon 79
  carp 103
  coefficients of variation in crosses 78–9
  differences between strains 76–9
  growth rates 76
  rainbow trout 77
  rainbow trout divergence 98

Sunbeams 172
Super-gene 42
Superoxide dismutase 61
Sutton, W. S. 110
Synbranchus bengalensis 181

Tetraploid 110, 206, 221–2
Thalassoma bifasciatus 182
Thunnus alalunga 88
Tiger trout 171
Tilapias 244–5
Tinca tinca 27
Tissue culture genetics 152, 232
Tissue grafting 186–7
Toothcarps 238–9
Transgenic *see* Gene transfer
Translocation 122
Triploid
  production 213–21
  recognition 213–16
  sexual maturation of 220
  uses of 221
  yield of 216–20
Tumour genes 63
Tuna blood groups 87

Umbra limi 152
Unisexual *see* Female-only
Unisexual triploids 187–8
Unisexuality 178

Vandellia cirrhosa 1
Vitellogenin 17

W chromosome 139–41, 155
Washington strain of rainbow trout 70
Wolffian ducts 181

X-linked inheritance 35
Xiphophorus helleri 28
Xiphophorus helleri × *X. maculatus* 63
Xiphophorus maculatus 30, 139

Y chromosome map of the Guppy 130
Y chromosomes in *Fundulus* 154
Y-linked inheritance 34–5
Yamamato, T. 193
YY males 144, 196, 213